D0275031

WENNER-GREN CENTER
INTERNATIONAL SYMPOSIUM SERIES

VOLUME 43

CENTRAL AND PERIPHERAL MECHANISMS OF COLOUR VISION

CENTRAL AND PERIPHERAL MECHANISMS OF COLOUR VISION

Proceedings of an International Symposium Held at The Wenner-Gren Center Stockholm, June 14-15, 1984

Edited by

David Ottoson
Professor of Physiology, Karolinska Institutet, S-10401 Stockholm, Sweden

and

Semir Zeki
Professor of Neurobiology, University College, London WC1E 6BT

First published 1985

Published by
THE MACMILLAN PRESS LTD
Houndmills, Basingstoke, Hampshire RG21 2XS
and London
Companies and representatives
throughout the world

Printed in Great Britain by
Camelot Press Ltd,
Southampton

British Library Cataloguing in Publication Data.
Central and peripheral mechanisms of colour vision.
—(Wenner-Gren Center international symposium series; v. 43)
1. Color vision
I. Ottoson, David II. Zeki, S. III. Series
591.1'823 QP483
ISBN 0-333-39321-X
ISSN 0083-7989

Contents

The Contributors

D.A. Baylor
Department of Neurobiology
Stanford University School of
  Medicine
Stanford
Connecticut 94305
USA

A. Blake
Department of Computer Science
University of Edinburgh
Edinburgh
UK

L.J. Friedman
IBM Corporation
Philadelphia
Pennsylvania
USA

J.P. Frisby
Department of Psychology
University of Sheffield
Sheffield S10 1ZN
UK

P. Gouras
Department of Ophthalmology
Columbia University
630 West 168 Street
New York 10032
USA

R. Granit
Eriksbergsgatan 14
S-114 30
Stockholm
Sweden

L.M. Hurvich
Department of Psychology
University of Pennsylvania
Philadelphia
Pennsylvania 19104
USA

D. Jameson
Department of Psychology
University of Pennsylvania
Philadelphia
Pennsylvania 19104
USA

K. Kirschfeld
Max-Planck-Institut für
  Biologische Kybernetik
Speemannstrasse 38
D-7400 Tubingen
Federal Republic of Germany

T.D. Lamb
Physiological Laboratory
Downing Street
Cambridge CB2 3EG
UK

E.H. Land
The Rowland Institute for Science Inc
100 Cambridge Parkway
Cambridge
Massachusetts 02142
USA

D.I.A. MacLeod
Psychology Department
University of California at San Diego
La Jolla
California 92093
USA

E. MacNichol
Marine Biological Laboratory
Woods Hole
Massachusetts 02543
USA

R. Menzel
Institut für Tierphysiologie und Angewandte Zoologie-Neurobiologie-Freie Universität Berlin
Königin-Luisestrasse 28-30
1000 Berlin 33
Federal Republic of Germany

C.R. Michael
Department of Physiology
Yale Medical School
333 Cedar Street
New Haven
Connecticut 06510
USA

B.J. Nunn
The Physiological Laboratory
University of Cambridge
Downing Street
Cambridge CB2 3EG
UK

E.N. Pugh Jr
Department of Psychology
University of Pennsylvania
Philadelphia
Pennsylvania 19104
USA

J.L. Schnapf
Department of Neurobiology
Stanford University School of Medicine
Stanford
Connecticut 94305
USA

J.E. Thornton
Polaroid Research Corporation
Cambridge
Massachusetts
USA

M.H. Yim
University of Texas
Health Sciences Center
Houston
Texas
USA

S. Zeki
Department of Anatomy
University College London
Gower Street
London WC1E 6BT
UK

E. Zrenner
Max-Planck Institute for Physiological and Clinical Research
Parkstrasse 1
D-6350 Bad Nauheim
Federal Republic of Germany

The Participants

Sten Sture Bergström
Department of applied Psychology
University of Umeå
S-901 87 UMEÅ
Sweden

Michael Burns
The Rowland Institute for Science
100 Cambridge Parkway
CAMBRIDGE
Massachusetts 02142
USA

Gunilla Derefeldt
National Defense Research Institute
Box 1165
S-581 11 LINKÖPING
Sweden

Walter Elenius
Department of Ophthalmology
University of Åbo
ÅBO 52
Finland

Curt von Euler
Department of Neurophysiology
Karolinska Institutet
S-104 01 STOCKHOLM
Sweden

Nicolas Franceschini
Department of Neurophysiology and Psychophysiology
C.N.R.S.
31 Chemin Joseph-Aiguier
F-13204 MARSEILLE
France

Kerstin Fredga
Swedish Board for Space Activities
Box 4006
S-171 04 SOLNA
Sweden

Christian Guld
Department of Neurophysiology
Blegdamsvej 3 C
DK-2200 KÖPENHAMN
Denmark

Karl-Arne Gustafsson
Department of applied Psychology
S-901 87 UMEÅ
Sweden

Anders Hedin
Department of Ophthalmology
Karolinska Hospital
S-104 01 STOCKHOLM
Sweden

Erik Ingelstam
Deaprtment of Optical Research
The Royal Institute of Technology
S-100 44 STOCKHOLM
Sweden

Craig Jason
The Rowland Institute for Science
100 Cambridge Parkway
CAMBRIDGE
Massachusetts 02142
USA

Gunnar Johansson
Döbelnsgatan 7
S-752 37 UPPSALA
Sweden

Olof Lagerlöf
Karlavägen 101
S-115 22 STOCKHOLM
Sweden

B. Lee
Max-Plank-Institut für bio-
physikalische Chemie
Postfach 2841
D-3400 GÖTTINGEN
FRG

Gunnar Lennerstrand
Department of Ophthalmology
Karolinska Hospital
S-104 01 STOCKHOLM
Sweden

Ivar lie
Department of Psychology
University of Oslo
P.O. Box 1094
OSLO 3
Norway

Sivert Lindström
Department of Physiology
University of Göteborg
S-400 33 GÖTEBORG
Sweden

Hans Marmolin
National Defense Research
Institute
Box 1165
S-581 11 LINKÖPING
Sweden

Dan Nielsen
Department of Neurophysiology
Blegdamsvej 3 C
DK-2200 KÖPENHAMN
Denmark

David Ottoson
Wenner-Gren Center
Sveavägen 166
S-113 46 STOCKHOLM
Sweden

Vagn Ousager Andersen
Department of Neurophysiology
Blegdamsvej 3 C
DK-2200 KÖPENHAMN
Denmark

Olle Pellmyr
Department of Entomology
University fo Uppsala
S-751 22 UPPSALA
Sweden

William Ray
The Rowland Institute for Science
100 Cambridge Parkway
CAMBRIDGE
Massachusetts 02142
USA

Tom Reuter
Department of Zoology
University of Helsinki
SF-00100 HELSINKI
Finland

Jyrki Rovamo
Department of Physiology
University of Helsinki
SF-00170 HELSINKI
Finland

Stuart Shipp
Department of Anatomy and
Embryology
University College London
Gower Street
LONDON WC1E 6BT
U.K.

Thorsten Seim
Department of Physics
University of Oslo
P.O. Box 1048
OSLO
Norway

Johan Sjöstrand
Department of Ophthalmology
Sahlgrenska Hospital
S-413 45 GÖTEBORG
Sweden

Carl Rudolf Skoglund
Brahegatan 58
S-114 37 STOCKHOLM
Sweden

Björn Tengroth
Department of Ophthalmology
Karolinska Hospital
S-104 01 STOCKHOLM
Sweden

Lillemor Wachtmeister
Department of Ophthalmology
Huddinge Hospital
S-141 86 HUDDINGE
Sweden

Arne Valberg
Department of Physics
University of Oslo
P.O. Box 1048
OSLO
Norway

Johannes Widakowich
Sturegatan 34
S-114 36 STOCKHOLM
Sweden

Torsten Wiesel
The Rockefeller University
1230 York Avenue
NEW YORK N.Y. 10021
USA

Birgitta Zetterström-Karpe
Department of Ophthalmology
Huddinge Hospital
S-141 86 HUDDINGE
Sweden

Preface

There seems little doubt that we are witnessing new and exciting developments in the field of colour vision and that these are beginning to change our views, not only of colour vision as a process, but also of the role of the brain in the perception of colours. Perhaps the two most important developments have been the somewhat belated realization of the magnitude of the problem that the nervous system has to solve in assigning colours to objects and the extent of the involvement of the cerebral cortex in this process. It seemed to us worthwhile to organize a small meeting to discuss these developments.

The standard meeting on colour vision revolves essentially around the physiology of the retina, with perhaps minor excursions into the lateral geniculate nucleus and, more recently, the cerebral cortex. It is usually complemented with psychophysical observations, based on wavelength matches in normal and abnormal subjects. We wanted to organize our meeting along a different line and emphasize the most fundamental property of the colour system, namely its ability to assign a (constant) colour to an object in spite of wide-ranging fluctuations in the energy and wavelength composition of the illuminant in which it is viewed. All the processes involved in colour vision must be working towards this end, without which colour vision would lose its significance as a meaningful signalling mechanism. The general question that the meeting addressed therefore was how the known physiology of the visual pathways, from the retina to central areas of the visual cortex, can account for this remarkable ability of the brain. We were very pleased that so many outstanding experts, representing different aspects of colour vision, could attend. We would, of course, have liked to invite many other participants to address this general question, but financial restrictions made this awkward while some of those we did invite were unfortunately unable to come.

It will become obvious, on reading these pages, that there are no ready answers but that there is an attempt to reconsider the known physiology, from retina to cortex, with this problem in mind. As with all new fields (and colour vision with this perspective may be considered to be a new field even if a great deal is known about it) there is still a struggle not only to define terms, but, more significantly, to formulate the correct questions which can then be experimentally tested. But we believe that these pages emphasize that to construct colours the nervous system has to solve a problem exceeding in complexity anything predicted by classical colour theories. This realization is, in itself, an achievement and if the book manages to communicate this to a wider audience, then our efforts will have been worthwhile.

Stockholm, June 1984

D. Ottoson
S. Zeki

Acknowledgements

The organizers wish to express their sincere thanks to Mrs Gun Hultgren and Miss Olga Popoff who, by their devotion and helpfulness, turned this meeting into such a pleasant event.

The meeting was made possible by the financial support of the Swedish Medical Research Council and the Wenner-Gren Center Foundation.

Introduction

R. Granit

Eriksbergsgatan 14, S-114 30 Stockholm, Sweden

David Ottoson will have to bear the full responsibility of setting up this Symposium on Colour Vision. I did not know of it at all before he kindly asked me to provide an Introduction.

Sixty years ago old Professor Robert Tigerstedt, then in Helsingfors, caught me doing some experiments in colour psychophysics and said: "Better give up that. All those who have taken to colour vision have failed to do anything sensible afterwards". In due course I transmitted this warning to William Rushton when he spent his sabbatical year in my laboratory in Stockholm in 1948. He then published it (1977), not mentioning Tigerstedt, in this version: "Long ago Granit warned me 'Colour is the femme fatale of vision. When once seduced, you will never be a free man again'. I was indeed seduced." However, this happened rather late in his life and so did not become catastrophic.

I, too was seduced in the mid thirties by work on the electroretinogram of the frog. In dark adaptation the blue sensitivity (450 nm) rose at a faster rate than that of rhodopsin at 500 nm. There were clearly blue-sensitive elements, very likely the green rods of Schwalbe, as since often confirmed. And, of course, the frog eye had a Purkinje shift, indicated already by Himstedt and Nagel in 1901 using a gas light spectrum. There was also Hecht's colour theory, irritatingly improbable in which wavelength differentiation was derived from small fringes of deviation from three otherwise wholly overlapping spectral sensitivity curves.

Keffer Hartline and I had been corresponding and so I was well aware of his success in isolating single fibres in the eye of the frog. Something had to be done about greater localization and isolation. But Hartline's technique was likely to require his own skill and patience. In the Preface to his collected papers (Hartline and Ratcliff, 1974) he said: "Successes were exasperatingly few, but they were welcome at the time."

There was a man I knew about in the neighbouring Anatomical Institute, Gunnar Svaetichin, who was trying to record by microelectrodes from cells in tissue cultures. He was quite willing to pull out some of these electrodes-- silver in glass-- for trial on the retina. For microstimulation I ordered special glass from Zeiss. They were rods that he pulled out into a fine tip and then silver-coated the surface so that the stimulating spectral light reached it by internal reflexion. Armed by two micromanipulators, one for the recording microelectrode, the other for the microstimulator, I went to work on the frog retina using the microscope light for light adaptation. At that time of his life Svaetichin neither took serious interest in vision, nor in the experimental labour but kindly delivered the things mentioned.

It was not difficult to match microelectrode and microstimulator. When, to begin with, the latter was held slightly above the retina, it produced a circular area of light that then had to be shrunk cautiously while one was trying to maintain spiking at the electrodes. Ultimately the tip of the microstimulator was pushed lightly into the retina and then the illuminated spot became virtually invisible. Sometimes a characteristic single-fibre spike was seen on the screen and followed acoustically, sometimes there was grouping of spikes. Often separation was then aided by the use of the threshold criterion for sensitivity to which only the most sensitive spike responded. Some help could be had by working on the turning point of the amplifier's grid curve to keep small spikes below noise level. Spectral energy distribution and stability were under full control. In light adaptation the thresholds were very precise.

At that time I had worked a great deal on light and dark adaptation of the frog eye and knew the temporal safety margin for the required light adaptation. Beginning dark adaptation was checked by the regular insertion of a reference wavelength as stimulus. There was often a considerable distance between the stimulating and recording points so that I concluded that the most likely origin of the spikes was a nerve fibre. Later, in Stockholm, I developed a coarse but much stabler platinum electrode.

Very few people would have read the original paper published in the Uppsala Läkareförenings Handlingar (1939) by Granit and Svaetichin. For this reason I have now detailed some essential technical points, described in still greater detail in the original paper. I have often seen the results misinterpreted. The so-called winter war interrupted further work in Helsingfors. In 1940 I was invited to Stockholm by the Karolinska Institutet.

The results of that early work were quite decisive on three fundamental points:

(I) Some spectral distribution curves of sensitivity were much narrower than that of rhodopsin, the only photochemical substance known at the time. R.J. Lythgoe had kindly given me his rhodopsin values.

(II) Others were broad, covering an area comparable with that of rhodopsin absorption.

(III) When a site that gave a broad curve with maximum around 560 nm was allowed to dark adapt, its sensitivity often-- though not always-- shifted to a new maximum around 500 nm. Clearly rods and cones projected to the same final common path.

I do not intend to review my old work on the retina during the war years in Stockholm on cone eyes, rod eyes, mixed mammalian eyes of cats, rats and guinea pigs. At that time the retina was something of a side line in world physiology and the psychophysicists disliked what was looking like a threat to their pet notion that the three König primaries would be deposited in the cortex unmodified. I concluded that I had done enough in the visual field of study. Counting my own psychophysical era I had in 1946 devoted a quarter century to it. I had ended in the grips of that grasping _femme fatale_ who would hold me to colour vision for ever. I was bored by the whole subject and well understood that a full stop was the only way of escape. I was also loaded with the responsibility of developing a neurophysiological centre in Scandinavia, having been provided with an Institute of my own for that purpose.

Work on the retina, including colour vision, needed fresh enthusiasm and no one can be more pleased than I to note the wave of new excitement that has been forthcoming, largely based on the gathering I now see in front of me. By your skill and imagination you have raised colour vision to a new level of actuality and it is indeed time for an overview. We are delighted to greet you here in Stockholm and are looking forward to a grand time of entertainment. You have succeeded in making the old _femme fatale_ respectable.

REFERENCES

Granit, R. and Svaetichin, G. (1939). Principles and technique of the electrophysiological analysis of colour reception with the aid of micro-electrodes. Uppsala Läkaref. Förh., 65, 161-177.

Hartline, H.K. and Ratliff, F. (1974). _Studies of Excitation and Inhibition in the Retina_. The Rockefeller University Press, New York.

Himstedt, F. and Nagel, W.A. (1901). Die Verteilung der Reizwerts für die Froschnetzhaut im Dispersionsspektrum des Gaslichtes, mittels der Aktionsstrome untersucht. Ber. naturf. Ges., Freiburg in

Br., _II_, 153-162.

Rushton, W.A.H. (1977). Informal essays on the history of physiology. In _Some memories of visual research in the past 50 years. The Pursuit of Nature_. Cambridge University Press, Cambridge.

Recent Advances in Retinex Theory

E.H.Land

The Rowland Institute for Science, Inc.,
100 Cambridge Parkway, Cambridge, Massachusetts 02142, USA

It is a cultural commonplace deriving from Newton that the colour of an object we see in the world around us depends on the relative amounts of red light, green light, and blue light coming from that object to our eyes. For a very long time it has been known that the colour of the object when it is part of a general scene will *not* change markedly with those considerable changes in the relative amounts of red, green, and blue light in the illumination which characterize sunlight versus blue skylight versus grey day versus tungsten light versus fluorescent light. This contradiction was named "colour constancy." Rather than dwelling on the explanations of colour constancy by Helmholtz and those who have followed him during the last century let us go on to show that the paradox does not really exist because it is not true that the colour of a point on an object is determined by the composition of the light coming from the object.

The first group of experiments is carried out with an arrangement of real fruits and vegetables (Figure 1).

> *In a dark room with black walls, three illuminating projectors with clear slides in the slide-holders are directed on the scene. The brightness of the projectors is individually controllable. An interference filter passing 450 nanometers is placed in front of one projector, 550 nanometers in front of the second, and 610 nanometers in front of the third. The measurements in this experiment are all carried out with a telescopic photometer that reads the flux towards the eye from a circular area about eight millimeters in diameter on the surface of an object. The readings are in watts per steradian per square meter. All readings are made with light from* **only one** *projector at a time. Three synchronized camera shutters on the illuminating projectors make possible the comparison of the colours in the fruit and vegatables as seen with continuous illumination and as seen with illumination for only a fraction of a second.*

The meter is directed at an orange, the fluxes towards the eye on the three

Figure 1. *Arrangement of Real Fruits and Vegetables*

wavebands are set equal to each other, and the orange is observed when the whole scene is illuminated with the combined light from the three illuminators. The orange is orange-coloured.

The process is repeated for a green pepper so that the identical radiation comes to the eye—the same three equal wattages as came from the orange. The green pepper is green.

Similarly, when the identical radiation reaches the eye from a yellow banana, the banana is yellow.

When the identical radiation reaches the eye from a dark red pepper, the pepper is dark and the pepper is red.

One of the most important experiments in this group is to compare the scene as viewed in continuous illumination with the scene viewed in a fractional second pulse. In view of the historic tendency to involve adaptation and eye-motion as causal factors in "colour constancy," it is most gratifying to see that for every new setting, as we turn our attention from object to object, the colours seen in a pulse are correct: Whether the scene is viewed in continuous illumination or in a fractional second pulse, straw is straw-colour, rye bread is rye-bread-colour, limes are lime-colour, green apples are green and red apples are red—with the illuminators of the scene so set in each case that the very same ratio of wattages of long, middle and short wave radiation comes to the eye from each of the objects in this list. That is, observed through a spectrophotometer, they would all look alike. Since this experiment establishes that the colour of an object is not a function of the composition of the light coming from the object, there is nothing surprising about the failure of the objects to change colour when the composition of the illumination changes: "Colour constancy" is the name of a non-existent paradox.

In Figure 1 you see a pile of orange-red pigment in front of the peppers. The pigment is red lead oxide, or minium, referred to by Newton (1704) in Proposition X, Problem V.

> "... Minium reflects the least refrangible or red-making rays most copiously, and thence appears red .... Every Body reflects the Rays of its own Colour more copiously than the rest, *and from their excess and predominance in the reflected Light has its Colour.*"

The demonstrations in this group of experiments prove that the part of Newton's proposition which we have italicized is incorrect. When the scene is so illuminated that the minium sends to the eye fluxes with the same three equal wattages, the minium continues to look its own brilliant orange-red colour—even though there is not at all an "excess and predominance in the reflected light" of "red-making rays."

This group of experiments leads to the first statement in Retinex Theory: **I. The composition of the light from an area in an image does not specify the colour of that area.**

* * * * *

When the fruit and vegetable scene is illuminated by light of one waveband, we observe that the very light objects stay very light and the very dark objects stay very dark as we alter the brightness of the illumination over nearly the whole range between extinction and maximum illumination. For example, with middle-wave illumination the red pepper will be always almost black and the green pepper always a light grey-green. This situation will be reversed when we change to long wave illumination; that is, the red pepper will always be light and the green pepper always black.

Based on these observations of the peppers, it is reasonable to venture the hypothesis that an object which always looks light with middle-wave illumination on the scene and always looks black with long-wave illumination on the scene will look green when the scene is illuminated with both illuminators and *will continue to look green* as we change the relative brightnesses of the two illuminators and hence the relative fluxes from the object to the eye. Similarly, an object which is black in middle-wave illumination and light in long-wave illumination will look red when the scene is illuminated with both illuminators and *will continue to look red* as we change the relative brightnesses of the two illuminators. Similar relationships can be established for the short-wave illumination.

If colour can indeed be predicted on the basis of the three lightnesses at a point, we are led to the question of how to predict *for each waveband separately,* each of the three lightnesses of a point in an image. More profoundly, since lightnesses themselves may not be an aspect of the colour, we would like to know how to compute the number on which lightness is based in the hope that we will find that a given trio of numbers will always be a single colour, a colour uniquely corresponding to the given trio. By learning how to compute the number associated with the lightness, we might be able to drop lightness itself out of consideration, using the computation only to give us a number which will immediately be associated with the two other numbers for that point, the trio of numbers being in effect the colour of that point.

These observations lead to the second statement of Retinex theory: **II. The colour of a unit area is determined by a trio of numbers each computed on a single waveband to give the relationship between the unit area and the rest of the unit areas on that waveband.**

* * * * *

As we contemplate the real still-life arrangement, still illuminating with one waveband, we note that the lightness of any given object does not change no matter where we put it in the whole arrangement, no matter what the background. If we put a neutral wedge into the slide holder of the illuminating projector so that the illumination transversely across the display is changing, for example, by a factor of ten from one side to the other, we find that the lightness of an object does not change as we move it from one side of the display to the other (Hering, 1920). The dark pepper is dark on either side; the light pepper is light on either side. This observation gives us pause because we were ready to suggest that the lightness of an object and of a unit area on it might be determined by the ratio of the flux from that unit area to the average of the fluxes from all the unit areas in the display. Clearly this simple hypothesis is threatened by the following facts: The ratio of flux per unit area from the pepper to the average flux per unit area from the whole field must change by a factor of ten as the pepper is moved across the field. Yet the *lightness* of the pepper is almost unchanged. Nevertheless, in spite of this significant difficulty, there is great appeal to the simple method for finding one of the three numbers for three wavebands, numbers which we hope will be our new designators for the colour of that unit area.

Since this simplest of approaches for the computation on a single waveband involves the ratio of the unit area to the average of all the unit areas in the field of view, let us note that there are two initially equivalent ways of stating the averaging process. The first is to take the ratio of the flux from the one unit area to the average of the flux from all the other unit areas. The second, which may seem somewhat esoteric, is to take the ratio of the unit area to each of the other unit areas in the field, separately, and then to average all of those ratios. We might say that in the first case we have established the relationship between the unit area and the average of *all* the other unit areas, whereas in the second case, we have found the average of all the *separate* relationships between the unit area and each of all the other unit areas. These two procedures yield identical results. The second one, however, lends itself to modification that will cause the computation to give the same value for the unit area when the overall illumination is altered to be mottled or uniform or oblique, whereas the first procedure will give variable results with such variation in the regularity of illumination.

The useful modification of the second procedure comprises noting any chain of unit areas from a distant one to the one we are characterizing and using the chain of ratios from the distant one to give the relationship to the distant one. It is required that when the individual ratios in the chain approximate unity, they act as if they *are* unity. With this procedure, in the presence of irregular illumination, the relationship of the unit area being evaluated to the distant unit area will yield the same number as the ratio of their fluxes in uniform illumination. If we now average a large number of ratios of our terminal unit area to many distant unit areas, we arrive at the number which characterizes that unit area for that waveband.

This second type of relationship is represented in the algorithm below which contains inherently all of the procedures necessary to satisfy the conditions to generate one number for the trio of numbers that will be the colour (Land, 1983).

Thus the third statement in Retinex theory is the algorithm itself.

III. **The Retinex algorithm:**

The relationship of i to j:

$$R^{\Lambda}(i,j) = \sum_k \delta \log \frac{I_{k+1}}{I_k}$$

$$\delta \log \frac{I_{k+1}}{I_k} = \begin{cases} \log \frac{I_{k+1}}{I_k} & \text{if } \left|\log \frac{I_{k+1}}{I_k}\right| > \text{threshold}, \\ 0 & \text{if } \left|\log \frac{I_{k+1}}{I_k}\right| < \text{threshold}. \end{cases}$$

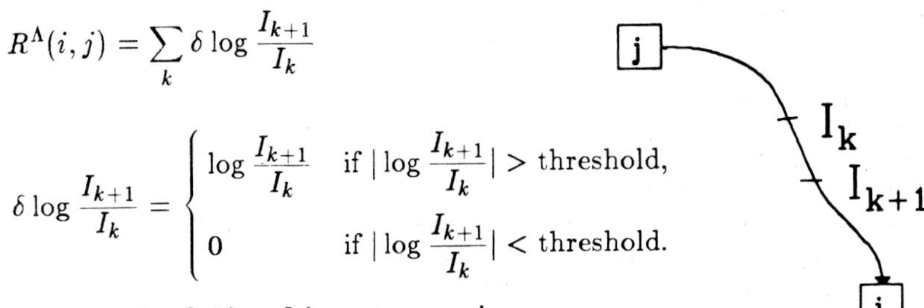

Average of relationships at area i:

$$\overline{R}^{\Lambda}(i) = \frac{\sum_{j=1}^{N} R^{\Lambda}(i,j)}{N}$$

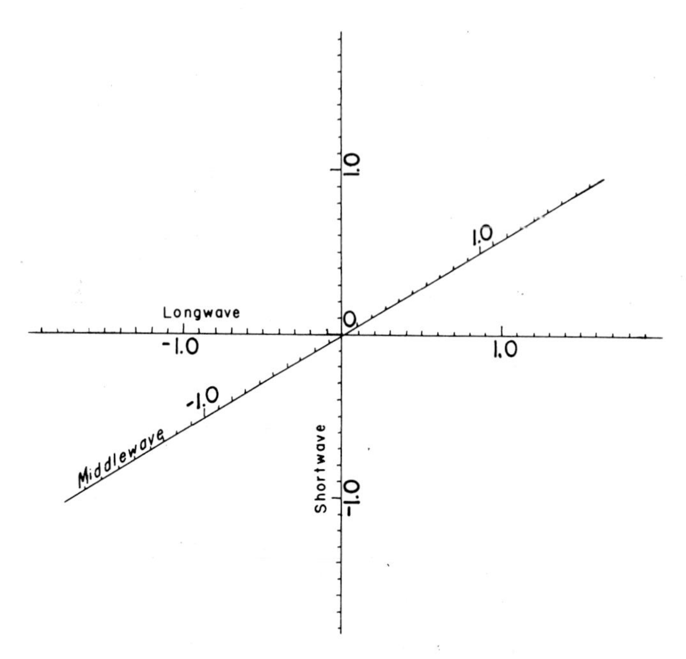

Figure 2. *Retinex 3-space Coordinate System*

The relationship of the flux at i to the flux at j, at one of the three principle wavebands, Λ, is given by the first two equations. The third equation gives the number which is one of the trio of numbers we have been looking for. Its value is the average of the relationships between unit area i and each of several hundred unit areas, j's, randomly distributed over the whole field of view. We have named this number *designator.* There will be three designators, one for each waveband. Together they can be represented as a point in a Retinex 3-space (Figure 2).

When an image of a coloured Mondrian (Figure 3), a collage of coloured rectangles, is formed on the retina and the computational technique proposed is pursued, it is found that the 3-space is populated in an orderly way. The points on one of the internal diagonals turn out to look black at one end, run through gray, and are white at the other end of the diagonal. There is a domain in which the greens reside, another for the reds, still another for the blues, and yet another for the yellows. It is a triumph of this computational technique that the overall variation in the composition of the illumination in terms of flux at a given wavelength or in terms of relative flux between wavelengths does not disturb the reliability with which a paper which looks red, no matter where it resides in the Mondrian, will have the same three designators as the other papers which look red. It will therefore be part of a family of reds which appears in one domain of the 3-space. Similarly, all the blues or greens or yellows, *wherever resident in the Mondrian* and however haplessly illuminated,

Figure 3. *Colour Mondrian*

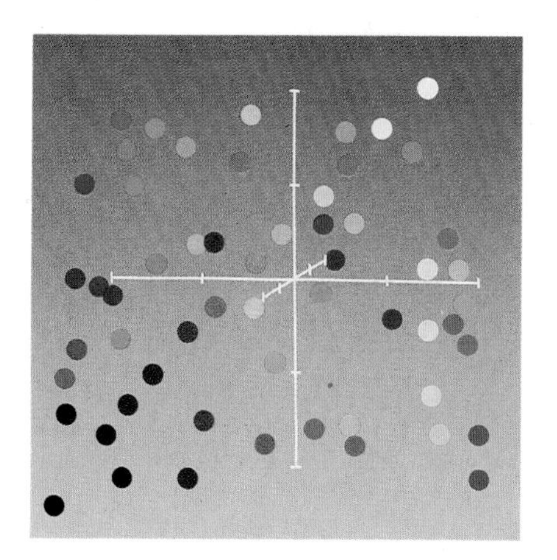

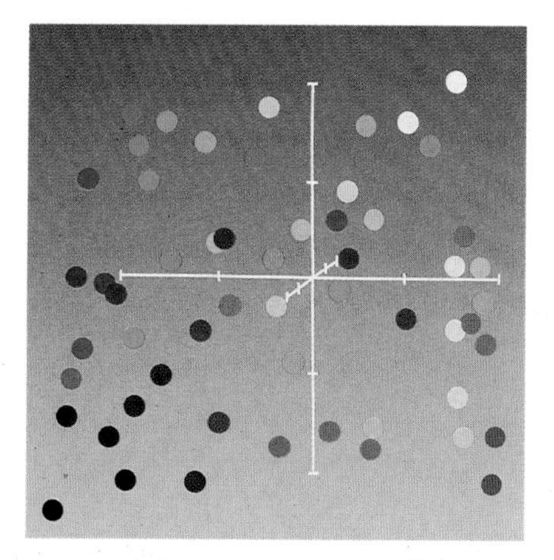

Figure 4. *Stereoscopic View of Retinex 3-space: left-eye view on the left.*

will appear in their appropriate domains in the 3-space (Figure 4). It is the computation that leads each paper to have its position in the 3-space; the proof of the pudding is that all things that appear in the same region of the 3-space are the same colour as one another, whatever their history in terms of geography and illumination on the Mondrian may have been (Land, 1983).

Measurements and Computations

While one family of measurements, such as those at the beginning of this paper, can be made with the real fruit scene, there are other measurements that can be carried out easily only with abstractions. The abstraction closest

Figure 5. *Collage of Coloured Objects*

to reality is provided by three black and white colour separation pictures of the fruit scene taken through the same three interference filters that we have been using for illuminating the live fruit scene. When we want the full-colour picture, the three black and white slides are projected in superposition, using the same three filters. Since all of the computations are carried out with reference to one waveband at a time, the measurements are for the most part made by projecting one or another of the three separations.

The next stage of abstraction (Land, 1964) uses a collage of coloured papers cut in the shapes of various objects (Figure 5). The objects are randomly arranged to satisfy the following four conditions:

(1) The objects overlap so that the support card does not show through gaps between them, in order to eliminate any common background.

(2) In spite of their overlapping, the nature of each object is immediately perceptible so that, for reliable discussion from a distance, it can be remembered and identified by name.

(3) Each object has in its immediate neighbors a large and unpredictable assortment of colours to eliminate any specifically coloured surround.

(4) Each object reflects not less than one-tenth of the light in those portions of the spectrum where the object has its least reflectance. This is to insure that surface light will be a small component of the measurement.

This collage can then be illuminated in exactly the way we illuminated the real fruit scene.

The third kind of abstraction is a collage of rectangular coloured papers

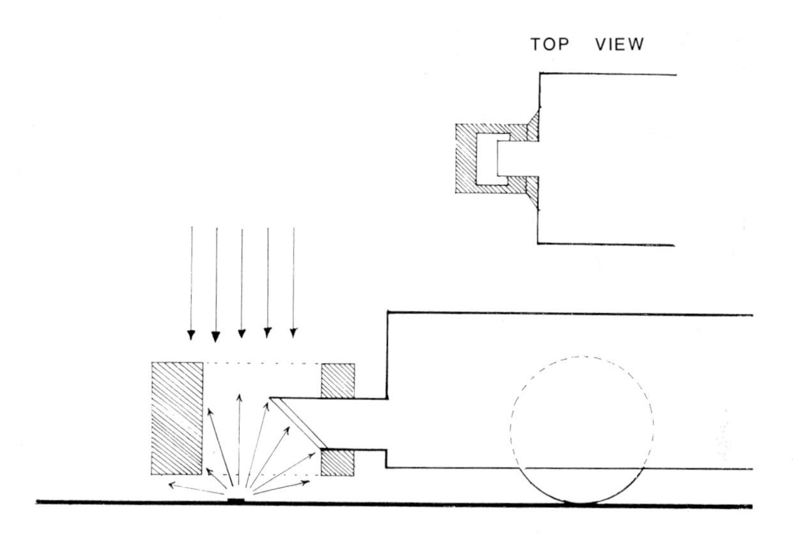

Figure 6. *Opti-Mouse: The rectangular aperture in top view (right) rests on the surface of the collage or screen. The projected radiation passes through the aperture to the surface where it is scattered back toward the (sloping) photosensitive element. A large number of rapid readings are integrated and read into the minicomputer every third of an inch of travel.*

assembled to look like a painting by Mondrian or Van Doesburg (Figure 3). At the time of our early experiments, it was difficult to devise satisfactory measuring and recording equipment to read a subject as intricate in detail as an average natural scene. Because the Mondrian is composed of extended uniform areas, it is easy to make measurements with the telephotometer and to assume continuity between any two measurements on any single rectangle on the Mondrian.

The Opti-Mouse

The relative simplicity of the algorithm as a formula for determining the colour of a unit area in any situation, natural or artificial, demanded that we design an instrument that could be swept across the surface of a display, whether projected, or planar like the real collages, to compute first the relationship between a unit area and any distant unit area (all on one waveband) and then, with the aid of a compact associated computer, the average of the relationships between the unit area and a large number of randomly selected unit areas throughout the field of view.

Our colleague, William Wray, designed the Opti-Mouse (Figures 6 and 7) and the associated circuits so that we are now able to carry out the algorithmic computation for a unit area in the Mondrian or the animal collage or the photograph of the fruit scene. What is particularly gratifying is to allow

Figure 7. *Opti-Mouse: bottom view (left), top view (right)*

the observer to see the display only in one waveband while the Opti-Mouse system is swept across the random pathways to give a number for that waveband for the unit area. The observer has not been allowed to see the display in full colour. Then the waveband is changed and the process is repeated for the same unit area. Finally, the number for the third waveband is computed. We now have the three designators. The 3-space, populated by a large number of previously computed colours, is projected in three dimensions onto a neighboring screen (Figure 4). The designators are located in the 3-space with the aid of a three-dimensional pointer. The colour where the three-dimensional pointer locates itself, as dictated by the three designators, is noted. Then and only then all three illuminators or projectors are turned on together to see if the colour predicted by the designators is indeed the colour of the unit area in the display. The predictions for the Mondrian and for the animal collage, both in uniform and graded illumination, have been strikingly good. The fruit scene is good with uniform illumination but for irregular illumination seems to require that the Opti-Mouse treat the ratio of adjacent readings as unity when the ratio is farther from unity than is necessary for the obliquely illuminated collages. [This suggests that, for the extreme intricacy of some parts of the pictures of real objects, the Opti-Mouse may need a smaller aperture and a closer spacing of adjacent readings than employed in these tests.]

Applications of Retinex theory and the Retinex algorithm

The predictive power of the algorithm is shown not merely by the examples we have discussed but also by many other experiments. For example, the colour of an area located on the Mondrian can be predicted and determined, while the flux to the eye is held constant, by modifying the designators within the area by computed changes in the rest of the field, for each waveband separately. Thus, the area can be changed from a specified white to a specified dark purple—without changing the flux to the eye from the area—by changes in the whole Mondrian computed for each waveband. In an important corollary experiment the area which is being changed from white to purple is surrounded by a very wide black border (produced by blocking the light at the slide-holder of the illuminating projector). The computed changes in the

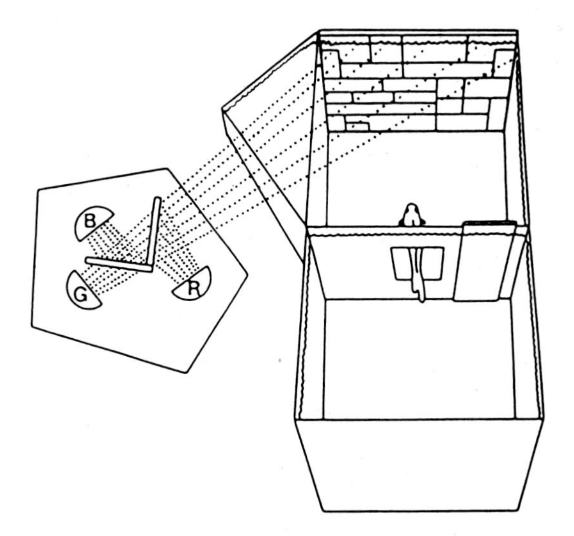

Figure 8. *Goldfish Entering Mondrian Chamber*

Mondrian, which had altered the designators for the area to make it white or purple, continue to be effective in the presence of the wide black border. This technique was employed as a tool, using an observer with a split brain, to show that the Retinex computation, wherever in the retinal-cortical system it is carried out, is long-range and requires cortical participation (Land, *et al.* 1983).

A most dramatic application of Retinex theory is the quantitative prediction of the colours available in the classical experiment of projecting two black and white pictures with red and white light to give a remarkable gamut of appropriate colours (Land, 1959). Furthermore, if the two projections of the red and white light are made in combination through a large long-wave transmitting filter and then through a large middle-wave transmitting filter and then through a large shor-wave transmitting filter, the Opti-Mouse, reading the image on each of the three wavebands separately, will correctly predict the three designators for locating the correct colours in the 3-space.

Professor Semir Zeki has discovered colour-reading cells in the V4 region of the prestriate visual cortex of the rhesus monkey (Zeki, 1980). The image of a Mondrian is formed on the retina of the anesthetized monkey, and Mondrian experiments analogous to those we carried out on ourselves are carried out with the animal. The results for the monkey, as reported by his cortical cells, are gratifyingly similar to our own.

Professor David Ingle, in our Laboratory, undertook the investigation of the question: "Is the goldfish a Retinex animal?" (Ingle, in press). The answer is dramatically in the affirmative: a goldfish selects a colour he ha been trained for when he is removed from the training tank and placed in a tank at one end of which is a Mondrian. In a series of tests, the Mondrian is so illuminated as to send to the fish's eye from an area of the colour he is trained for the same number of photons of the same frequency as came from each of many other colours in the Mondrian under some standard illumination (Figure 8).

We shall not undertake to tell here the other important techniques of computing Retinex relationships, particularly the recent modification by Blake (this volume) of Horn's (1974) formulation. Retaining all of the concepts in the Retinex algorithm, Blake can replace the one-dimensional computation by a two-dimensional computation. Finally we shall not examine here the coordinate transformations that take Retinex theory into the domain of double-opponent cells in the visual pathway (Livingstone and Hubel, 1984 and Land, 1983).

In summary the three propositions of Retinex theory are

I. **The composition of the light from an area in an image does not specify the colour of that area.**

II. **The colour of a unit area is determined by a trio of numbers each computed on a single waveband to give the relationship between the unit area and the rest of the unit areas on that waveband.**

III. **The trio of numbers, $\overline{R}^{\Lambda}$, as computed by the Retinex algorithm designate the point in Retinex 3-space which is the colour of the unit area.**

References

Hering, E. (**1920**). *Grundzüge der Lehre vom Lichtsinn.* Springer, Berlin, 30.

Horn, B.K.P. (**1974**). *Determining Lightness from an Image.* Computer Graphics and Image Processing, 3, 277-299.

Ingle, D.J. (in press). *The Goldfish is a Retinex Animal.* Science.

Land, E.H. (**1959**). *Color Vision and the Natural Image. Parts I and II*, Proceedings of the National Academy of Sciences, 45, 115-129 and 636-644.

Land, E.H. (**1964**). *The Retinex.* American Scientist, 52, 247-264.

Land, E.H. (**1983**). *Recent Advances in Retinex Theory and Some Implications for Cortical Computations: Color Vision and the Natural Image.*

Proceedings of the National Academy of Sciences, 80, 5163-5169, and references therein.

Land, E.H., Hubel, D.H., Livingstone, M.S., Perry, S.H. and Burns, M.M. (**1983**). *Colour-Generating Interactions Across the Corpus Callosum*, Nature, 303, 616-618.

Livingstone, M.S. and Hubel, D.H. (**1984**). *Anatomy and Physiology of a Color System in the Primate Visual Cortex.* Journal of Neuroscience, 4, 309-356.

Newton, I. (**1704**). *Opticks* (London) Prop. X, Prob. V, 135.

Zeki, S. (**1980**). *The Representation of Colours in the Cerebral Cortex.* Nature, 284, 412-418.

Colour Pathways and Hierarchies in the Cerebral Cortex

S. Zeki

Department of Anatomy, University College London, Gower Street, London WC1E 6BT, UK

Introduction

My interest in colour vision is a direct consequence of my studies on the anatomical and functional organization of the prestriate cortex which, in the macaque monkey, is a large strip of cytoarchitectonically uniform cortex surrounding the striate cortex (V1). Long regarded as a single cortical field, it was commonly referred to as the "visuo-psychic" band or the "visual association" cortex with the implication that, unlike V1 or the "visuo-sensory" cortex (Campbell 1905), which was thought to be involved in visual sensation, the prestriate cortex was more involved in visual perception. The difficulty of demonstrating any clear and obvious visual defect after lesions of the prestriate visual cortex (see Zeki 1969 for a review), compared to the obvious scotomas which obtain after V1 lesions, served naturally to reinforce this view.

A twist in this view of the organization of the visual cortex was introduced by the anatomical demonstration that the prestriate cortex, far from being the single cortical area which its uniform cytoarchitecture had for so long implied, is in fact composed of a multiplicity of distinct visual areas, each with its own distinct input from V1 and its own set of callosal connections (Cragg 1969, Zeki 1969, 1970, 1971) (Figure 1). Parallel physiological studies indicated that different visual areas execute different visual functions (Zeki 1973, 1974), a finding which led to the theory of functional specialization in the visual cortex (Zeki 1978 a & b). This is a modern re-statement of the theory of functional localization (von Economo and Koskinas 1925) which clinical scientists had sought to establish, but which came to be doubted, not only because of the imprecise nature of clinical defects, frequently associated with other, and more general, disturbances but

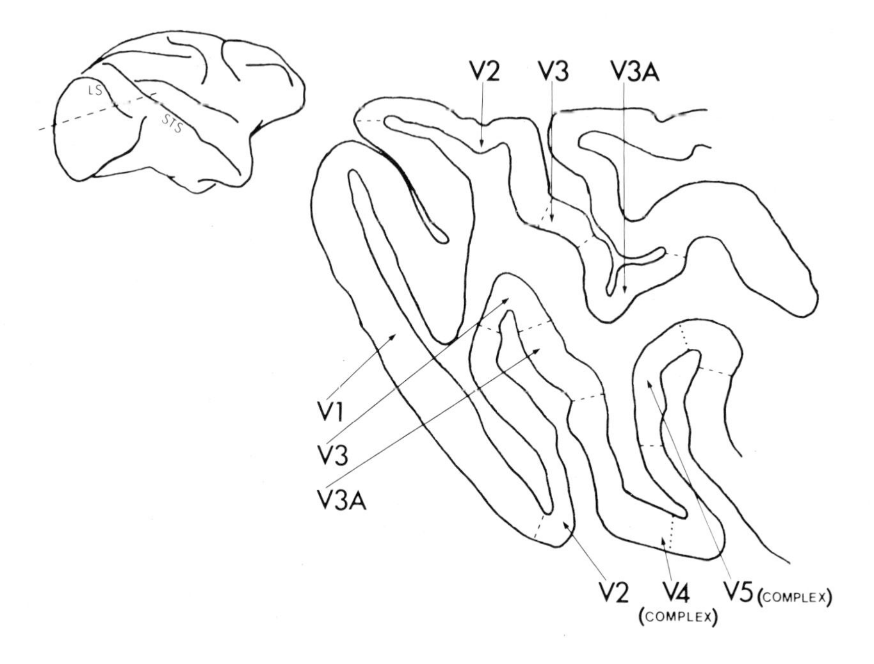

Figure 1: A drawing of a horizontal section through the brain of the macaque monkey, taken at the level indicated, to show the disposition of the various prestriate areas. V5 (the motion area) receives a direct input from V1 but is part of a larger complex, the V5 complex, the rest of which does not receive a direct input from V1. There are at least two, and possibly more, areas within the V4 complex. LS = Lunate Sulcus, STS = Superior Temporal Sulcus.

also because of the difficulty of reproducing such defects experimentally with lesions of the prestriate cortex.

Functional specialization and the serial doctrine.

The demonstration that there are other visual areas besides the striate cortex and that there may be more than one such extrastriate visual area had, in fact, been shown in other animals and had been suspected ever since Campbell (1905) had stated, with Delphic wisdom, that the prestriate cortex may contain "one or more" areas. Indeed, Clare and Bishop (1954) had shown that, in the cat, there is a visual area remote from the striate cortex and Hubel and Wiesel (1969) and Wright (1969) had not only recorded from this area, with its heavy concentration of directionally selective cells, but also from two further areas, 18 & 19 (Hubel and Wiesel 1965), which had previously been defined anatomically by Otsuka and Hassler (1962). But the interpretation given to the results of these studies was different from mine. Hubel and Wiesel had supposed from their studies that each visual area was analyzing all the visual information but at a more complex level than the antecedent areas. Thus areas 18 & 19 of the cat were thought to involve higher order analyses of form (Hubel and Wiesel 1965) and they were puzzled at what further processing could occur in the Clare-Bishop area (Hubel and Wiesel 1969). Their results led them to the conclusion that these areas operate in hierarchical fashion, with cells at more central areas assembling the information analyzed in more antecedent areas, in "building block" fashion, a notion almost certainly derived from the structuralist doctrine of the psychologists (Hochberg 1964). This supposes that just as a percept can be broken down into its components by the process of introspective psychology, so the nervous system builds up an image of the external world by analyzing it point by point and then assembling this information in piecemeal fashion. Our view, by contrast, was that different visual areas process different kinds of visual information. For example, my interpretation of the results of recordings from the motion area (V5), which also has a high concentration of directionally selective cells (Zeki 1974), was not that it is involved in a more complex analysis of the visual scene than antecedent areas, but that it is involved in a different kind of visual processing, namely the analysis of motion.

There is little doubt that a cardinal factor, but not the only one, which led to an alternative view of the organization of cerebral processes involved in vision was colour. My work was almost solely concerned with the macaque monkey, an animal close to man and, like him, possessing good colour vision (Morgan 1966; De Valois et al. 1974). The original observation (Zeki 1974) that a well defined visual area, V5, in this animal has clearly driven visual cells, most of which are directionally selective and none of which is colour or wavelength selective (see also Gattass and Gross 1981; Van Essen et al. 1981), was as impressive as the observation that there are other visual areas, such as those comprising the V4 complex, in which there are heavy concentrations of wavelength or colour cells (Zeki 1974, 1975, 1977, 1983; Andersen et al. 1983; Tanaka et al. 1984). It would have been difficult to obtain so clear-cut a result in the cat, an animal with a somewhat primitive colour system, and thus reach the conclusion that some areas are dealing with colour whereas others not. We were also impressed by finding that areas such as V3 and V3A also had virtually no colour or wavelength selective cells (Zeki 1978 a & b; Baizer 1982; Andersen et al. 1982). Again, such evidence is easier to obtain in the monkey than in the cat. It was thus both the positive and negative evidence that led to the concept of functional specialization. But although such a concept has logical consequences, outlined below, one must not view it as an alternative to the hierarchical doctrine but rather as a more general feature of the organization of the visual areas of the cerebral cortex, into which the hierarchical concept must be

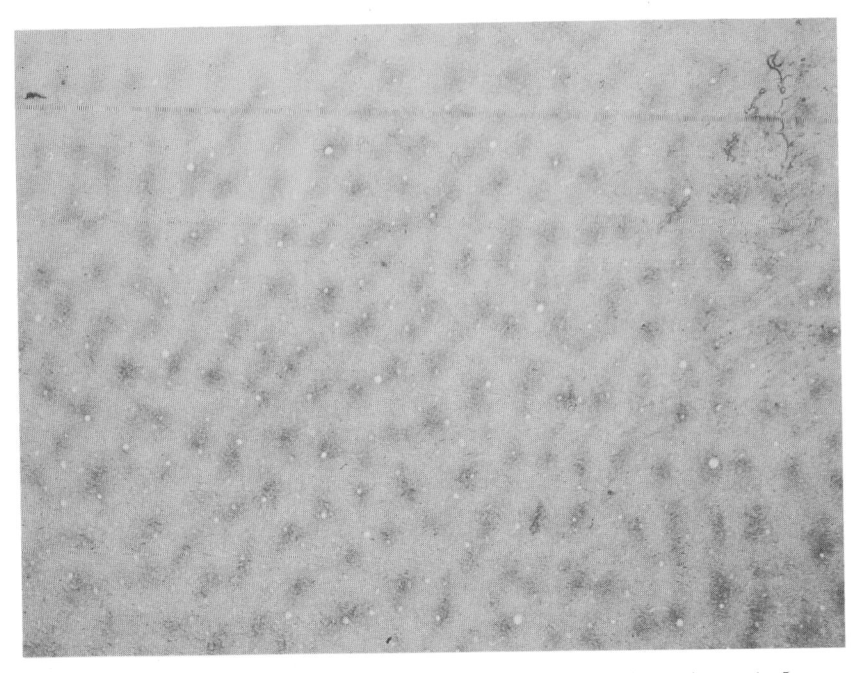

Figure 2: Photomicrograph of a tangential section through layers II and III of macaque monkey striate cortex (V1), stained for cytochrome oxidase.

incorporated. Put more simply, we view each functionally distinct pathway as incorporating a set of hierarchies and nowhere is this more evident than in the colour pathways within the cerebral cortex.

Functional specialization and the organization of the striate cortex (V1).

Since most of the functionally distinct visual areas receive their input from V1, one logical consequence following from the demonstration of functional specialization within the prestriate cortex is that V1, which receives the predominant input from the eyes through the lateral geniculate nucleus, must act as a segregator of information, parcelling out different kinds of information to the different visual areas for further processing (Zeki 1975). This should be so because it would be difficult to imagine that V1 would send the same information to these different prestriate visual areas in these independent and parallel pathways. At the time, direct evidence for such a role for V1 was not available. In spite of evidence which showed that there must be high concentrations of wavelength selective cells in V1, especially parts of it representing central vision (Poggio et al.

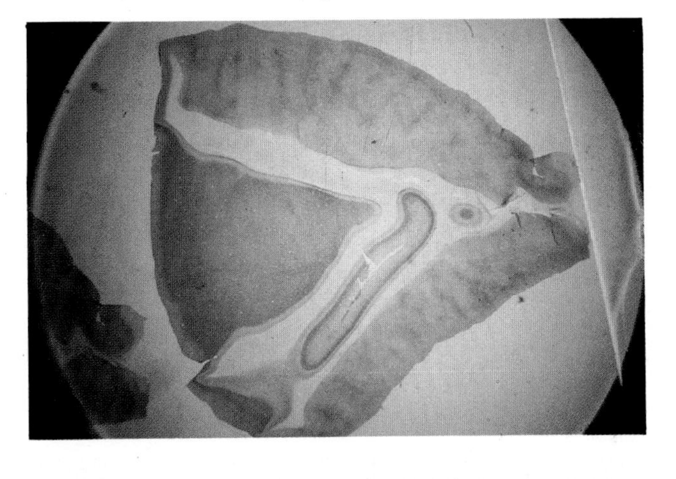

(A)

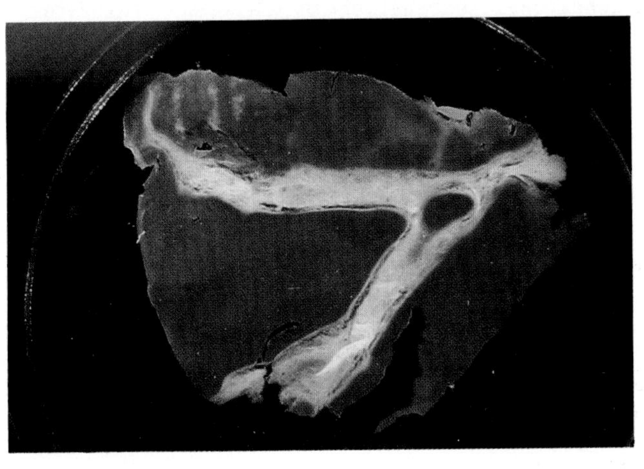

(B)

Figure 3: (A) Photomicrograph of a section through the flat-mounted occipital operculum of a macaque monkey, stained for cytochrome oxidase. Note the striped distribution of regions of high cytochrome oxidase content with V2. (B) Dark field photomicrograph of a section through the flat-mounted occipital operculum of a macaque monkey in which wheat germ agglutinin HRP was injected into V4. Note the distribution of the HRP label in stripes within V2, similar in disposition to the cytochrome oxidase stripes seen in A.

1975; Zeki 1983) and that such cells may be grouped together and separated from orientation selective but wavelength non-selective cells (Michael 1981), the most complete and detailed studies of the functional architecture of V1 had shown that all cells outside layer 1VC were orientation selective (Hubel and Wiesel 1974, 1977), including the cells of layers II & III which provide a major output to the visual areas of extrastriate cortex. But the development of cytochrome oxidase histochemistry (Wong-Riley 1979), coupled to electrophysiology (Horton and Hubel 1980), has modified our picture of the functional architecture of V1. It now appears that, in spite of its striking cytoarchitectural uniformity, V1, and especially its upper two layers, are made up of discrete 'blobs' of relatively high cytochrome oxidase content which, when viewed in tangential section, takes on a polka dot pattern (Figure 2). The studies of Livingstone and Hubel (1984) have shown that the double-opponent 'colour coded' cells (see below) most of which are not orientation selective (but see Michael 1978 and also this volume for an opposite result) are concentrated within the blobs whereas regions in between the blobs (the inter-blob zones) contain predominantly orientation selective cells, most of which are not wavelength specific.

This direct anatomical and physiological evidence shows that there is a functional segregation in the tangential direction within layers II & III of V1 and is a striking confirmation of the conclusion, reached from a study of the outputs from V1 to different prestriate visual areas, that V1 acts as a segregator of different kinds of visual information. Anatomical studies show further that there is another, radial, segregation within V1 since the cells projecting to V5 in the superior temporal sulcus are located within layer IV and, more infrequently, within the upper part of layer VI whereas those projecting to V2 are located in layer II and III (Lund et al. 1975; Shipp and Zeki 1985). It is almost certain that more evidence in favour of this segregation will accrue over the coming years. But the evidence presently available establishes that V1 segregates different kinds of visual information, at least insofar as colour, form and motion are concerned. In a sense the polka dot cytochrome oxidase pattern of V1 is, to date, the most eloquent testimony to this important role of V1

Separate colour and motion pathways.

Another logical consequence of functional specialization is that the pathways dealing with different kinds of visual information are themselves separate and the available evidence shows that this is indeed so. It is known that V1 projects both to V2 and V5, among other visual areas (Cragg 1969; Zeki 1969, 1971 a & b, 1978c, 1981; Ungerleider and Mishkin 1981; Van Essen et al. 1981). V2, in turn, projects to V5 as well as to V4 (Zeki 1971, 1976). Because of the profound difference in the functional properties of V5 and the V4 complex, it is natural to suppose that the pathways to these two areas are maintained separate within V2 and that the latter, like V1, must act as a segregator of information. That this is so is suggested by the demonstration that injections of wheat germ agglutinin horse radish peroxidase (HRP) into V5 or into V4 shows that neurons projecting to either area are not continuously distributed within V2 but occur in clusters

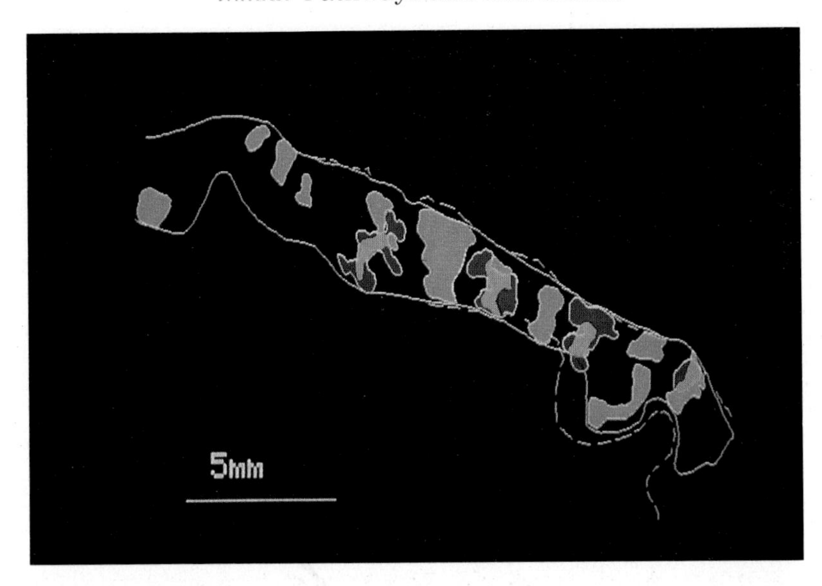

Figure 4A: Computer aided reconstruction of the pattern of distribution of cytochrome oxidase and HRP label from 2 adjacent sections in V2 (posterior bank of the lunate sulcus) in an animal in which HRP was injected into V4. Areas of labelled cells are shown in red, of high cytochrome oxidase content in green and regions of overlap between the two in yellow. Note that labelled cells are centered around alternate regions of high cytochrome oxidase content, apparently the thinner set.

(Fries and Zeki 1983; Shipp and Zeki 1984). When V2 is sectioned tangentially, or when it is reconstructed from horizontal sections to show the tangential distribution of labelled cells, it becomes evident that the distribution of labelled cells in V2, following injections into either area, is in the form of stripes which run orthogonal to the V1-V2 boundary (see Figure 3). The configuration of the stripes of labelled cells within V2 is remarkably similar to that of the cytochrome oxidase stripes within it. We have therefore tried to relate the distribution of cells in V2 projecting to either area with the pattern of cytochrome oxidase stripes in it. Figure 4 is a computer aided reconstruction, made from horizontal sections, to show the tangential distribution of labelled cells following injections of HRP label into V4 and into V5. The remarkable result is that labelled cells following injections made into either area are centered around the cytochrome oxidase stripes, with the interstripe zones being largely free of such cells. As striking is the fact that, following injections into either area, labelled cells are not found in every cytochrome oxidase stripe but in every other one. This evidence suggests (a) that the pathways leading to V5 and to V4 are centered around regions of high metabolic activity in V2,

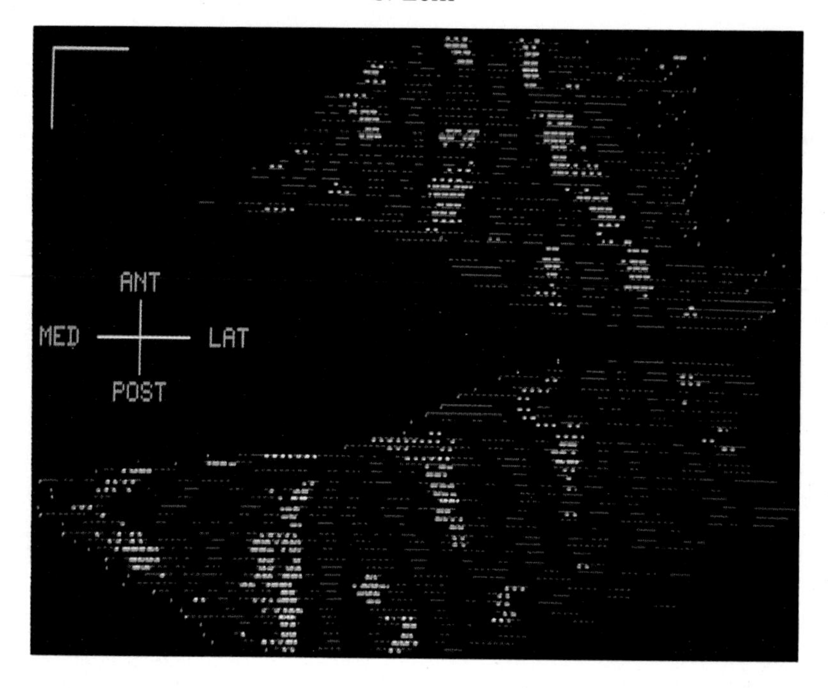

Figure 4B: A computer aided reconstruction of the distribution of cells projecting to V5 (red dots) and of regions staining densely for cytochrome oxidase (grey lines) in area V2 of macaque visual cortex (HRP label was injected into V5). The reconstruction is symmetrical about the fundus of the inferior occipital sulcus - a part of V2 representing upper visual field - and mimics the appearance that might be obtained if the banks of the sulcus were flattened. Each line represents information from a single section through the sulcus, and these have been aligned with reference to cortical landmarks. The cytochrome oxidase pattern is not clear in all places, but the familiar picture of alternate thick and thin bands can be seen to emerge. The output to V5 arises from alternate cytochrome oxidase dense bands, seemingly the thicker ones. Scale bar is 1.7 mm for both axes.

for reasons yet to be determined; (b) that the pathways leading to the two areas are separate, with alternate cytochrome oxidase stripes being concerned with relaying different kinds of information. The critical experiment to prove (b) has yet to be performed; it would involve injecting separate anatomical lables into V4 and V5 and correlating the distribution of the two labels with the cytochrome oxidase stripes in V2. But, given the separation of pathways within V1, it would be surprising if such a separation were not maintained within V2 and the evidence given above suggests that this is a strong possibility. If so, it would imply that there are two functionally distinct cytochrome oxidase stripes, one relating to colour and the other to motion. Whether this is so and whether there is any correlation between such possible groupings and the thick and thin stripes remains to be determined. The apparent separation of motion and colour pathways within V2 gains added significance from the results of Livingstone and Hubel (1982) who have shown that the blobs in V1 project to the stripes of V2 and the interblob regions to the interstripe zones. It thus appears that where a well defined area, such as V1 and V2, contains a variety of cells, the different types of cells are maintained segregated within it, a segregation which is over and above the topographic representation.

A third expectation from the demonstration of functional specialization is that lesions in specific areas of the prestriate cortex should lead, not to global scotomas, but to perceptual scotomas in specific submodalities of vision. Indeed, human studies had indicated that, following lesions in extrastriate cortex, colour vision could be specifically affected (see Meadows 1974 for a review). Recently Wurtz and his colleagues have shown that following small, chemical, lesions in V5, the monkey's ability to perceive moving objects is impaired. The effect is transient but statistically significant. As well, Wild et al. (1984) have shown that, following lesions of the V4 complex, monkeys are significantly impaired in their ability to discriminate the colour of objects when the ratio of incident illumination changes. Such lesions do not affect the monkeys' ability to discriminate orientations and their ability to discriminiate wavelengths as such is only very mildly impaired (Fries and Zeki 1981), an observation that acquires added significance from physiological results described below. This evidence from monkeys thus complements evidence from clinical studies which show that motion and colour discrimination can be specifically impaired following lesions involving extrastriate cortex in man (Zihl et al. 1983; Pearlman et al. 1976; Damassio et al. 1980).

Functional specialization and brain algorithms.

It is almost certain that not all the visual areas of the prestriate cortex have been charted and that more remain to be discovered. Nevertheless, the evidence to date is sufficient to establish (a) the principle of the multiplicity of visual areas within prestriate cortex; (b) the functional specialization of different groups of areas for processing different kinds of information and (c) the segregation of pathways dealing with different aspects of the visual scene. It thus becomes important to ask why it is that the cortex uses this strategy of parcellation. It has been suggested (Zeki 1981) that different visual areas are involved in constructing different kinds of visual experience and that the requirements for constructing each are sufficiently different for the cortex to use different visual areas, with different internal wiring patterns and connections, to construct each category. Perhaps the simplest example is that of colour vision.

Figure 5: Natural colours and void colours. When the green area of the multicoloured display on the left is made to reflect 60, 30 and 10 mW Sr^{-1} m^{-1} of long, middle and short wave light and viewed as part of the complex scene, it looks green; when, reflecting the same energies, it is viewed on its own, by excluding other areas from the field of view, it appears a light grey. (Based on Land, 1974).

Land (see this voloume and Land 1974, 1983) has shown in a rigid quantitative way that the colour of a surface is determined not only by the wavelength composition of the light reflected from that surface but from all other surfaces in the field of view (I use the term colour to mean colour as perceived by a normal human observer). Land's experiments present an important departure from traditional psychophysical experiments which have been more concerned with what determines the colour of a single region in the field of view, to the exclusion of all else (the reduction screen or "void" experimental situation, see Figure 5). If the strategy that the nervous system uses to determine the natural colour of surfaces is at all similar to the algorithms postulated by Land and others, it follows that a 'point-by-point' analysis of wavelength followed by the assembling of the points so analyzed, in building block fashion, to determine the colour of surfaces, would be insufficient to construct colours. That the nervous system samples the wavelength composition coming from many different areas to determine the colour of each essentially makes it independent of the precise wavelength composition coming from one area alone. This leads to the most striking property of the colour system, namely that the colour of an object or surface which forms part of a complex multicoloured scene changes very little, if at all, with profound changes in the wavelength composition of the light reflected from it. This phenomenon is commonly referred to as colour constancy. Had it not been for it, colour vision would lose its significance as a biological signalling mechanism. For if the colour of objects were to change with every change in the wavelength composition of the light reflected from them, then the presence of objects would no longer be signalled by their colour. The problem of how the nervous system assigns a constant colour to a surface, in spite of wide ranging changes in the energy and wavelength composition of the light reflected from it, thus becomes the central problem of colour vision. We have discussed the relationship of physiological findings to Land's computational scheme elsewhere (Zeki 1983 a & b, Zeki and Shipp, 1985). Even though Land's retinex theory is not physiologically based but more in the nature of a computational scheme which tries to account for how the reflectance of surfaces can be generated from the wavelength intensity information reaching the organism, it is nevertheless interesting to consider it in a physiological context because, whatever the nature of the computations used by the brain to generate colours, the end result is the same as predicted by retinex theory.

Algorithms based on retinex theory have been proposed (Land 1983 and this volume; Horn (1974); Blake (this volume)). These share common principles but differ in the details of implementation. Each involves a local differencing stage in which the reflectance, at a given waveband, of neighbouring points along paths leading from the entire field of view to a unit area X which forms part of a multicoloured scene, is determined. A thresholding operation eliminate differences below some arbitrary threshold, due to oblique or uneven illumination, and includes larger ones in the calculation. These local operations lead to a global one in which the average relative reflectance of the surround with respect to X is determined for each of the three wavebands. The colour of a surface is then determined by a comparison of the values achieved through each waveband.

The double opponent cell as a lightness generating system

The minimum neural requirement for implementation of the Land algorithm would be a cell which is able to compare the amount of light of a given waveband, say long wave, reflected from one part of its receptive field and light of the same waveband reflected from adjacent parts. The simplest such cell, first described by Nigel Daw (1967), is known as the double opponent cell. In the monkey, these cells are not found in the retina or lateral geniculate nucleus, but common in striate cortex (Wiesel and Hubel 1966; Michael 1978 a & b and this volume; Livingstone and Hubel 1984). A common kind of double opponent cell is one which gives an ON response to long-wave light and an OFF response to middle wave light in the centre of its receptive field and an opposite kind of response in the surround. I have studied the reaction of

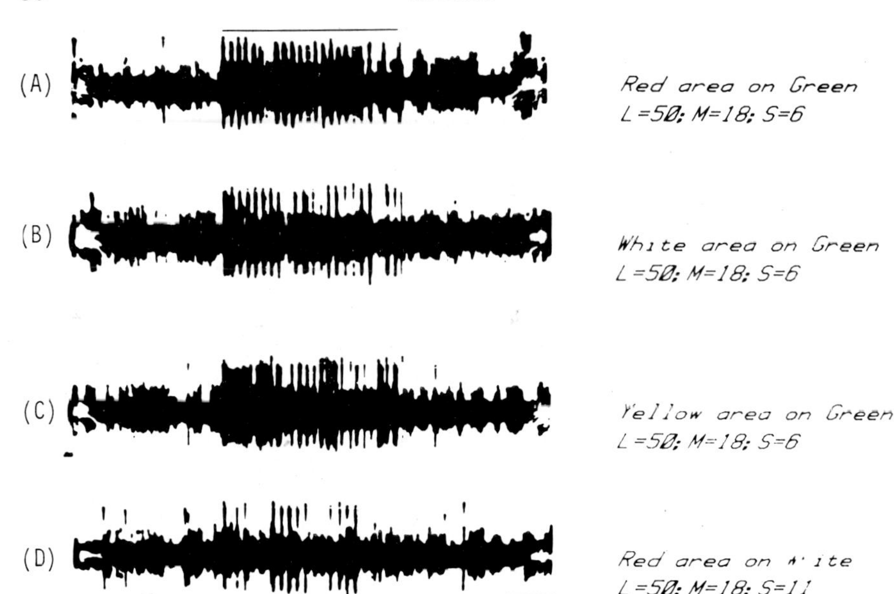

Figure 6: The responses of a double opponent cell (long wave on, middle wave off centre) in monkey striate cortex to areas of different colour. In each, the area was made to reflect 50, 18 and 6 mW Sr^{-1} m^{-2} of long, middle and short wave light and, in traces A-C, moved into the cell's receptive field centre against a green background. In traces A-C the green background was reflecting in A, 8.0, 35, and 7; in B, 8.0, 9, 6 and in C, 8, 10, 7 mW Sr^{-1} m^{-2} of long middle and short wave light respectively. In trace D, the red area still reflecting the same energies was placed against a white background, which was reflecting 63, 135 and 28 mW Sr^{-1} m^{-2} of long, middle and short wave light respectively.

such cells to colours, using an experimental condition similar to that of Land's (Zeki 1983a). Figure 6 shows the reaction of a long wave ON middle wave OFF centre double opponent cell to areas of different colour when put against different coloured backgrounds. It turns out that, to respond, the cell requires a minimum difference in the amount of long and middle wave light reflected from its centre and surround. Within these limits, it is indifferent to the colour of the area in its receptive field. Thus the cell of Figure 6 gave a good response to a red, white, yellow, light grey or brown area in a green surround when each area was reflecting 50, 18 and 6 mW. Sr^{-1} m^{-2} of long, middle and short wave light. Under these conditions, the green surround was obviously reflecting much less long wave light. In each case, a neutral point could be found when the difference in the amount of long and middle wave light reflected from centre and surround was such that the cell did not respond, even though with such

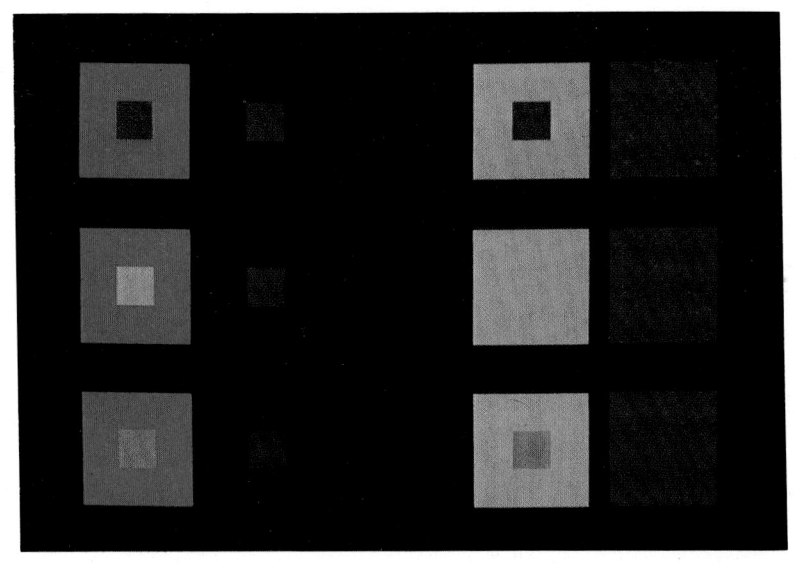

Figure 7: Lightnesses of red, white and yellow areas against a green background when the scene is illuminated with long wave light alone (left hand columns). The right hand columns show the lightnesses when the areas are placed against a white background and viewed in long wave light. The responses of double opponent cells correlate with lightnesses, not colours but whether they correlate with two sets of lightnesses simultaneously (e.g. long and middle wave lightnesses) depends upon the strength of long and middle wave inputs to centre and surround.

variations the colours of the areas did not change. On the other hand, if a white surround was substituted for the green one, the cell's response was reduced. Thus the responses of this and other, similar, double opponent cells do not correlate obviously with colours. We suggest, therefore, that the double opponent cell may be the first stage in local wavelength differencing and thus the generation of lightnesses. Consider the reaction of the cell of Figure 6 to a red, yellow and white area in a green surround (see Figure 7). In each case when this scene is viewed in long wave light or through a long wave filter, the centre (which has a higher reflectance for long wave light) appears light compared to the surround. Thus the common feature in all these stimuli is a difference in reflectance for long wave light (and therefore lightness) between centre and surround. On the other hand, when each area is put against a white background and viewed in long wave light, the entire scene is light (because white has a high reflectance for light of all wavebands). Thus the responses of these cells correlate well with reflectances for long wave light, but not with colour (whether the responses of such a double opponent cells also correlate with lightness produced by middle

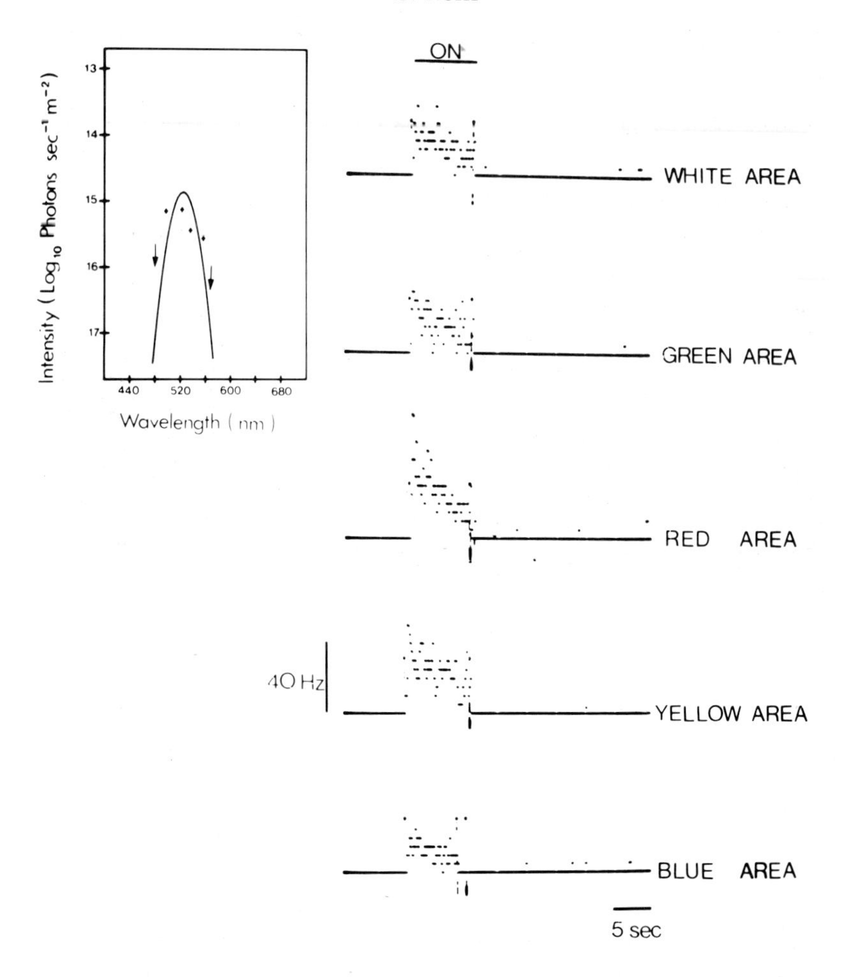

Figure 8: The responses of a middle wave selective cell in V1 to areas of different colour (see inset for its action spectrum). The responses of the cell are shown as its discharge frequency. Each area when placed in the receptive field of the cell was made to reflect 69, 33 and 7 mW Sr^{-1} m^{-2} of long, middle and short wavelength light respectively. The receptive field size of this binocularly driven cell was 1° x 1° and situated within the central 1°. It was stimulated through the contralateral eye alone. In the action spectrum, downward arrows indicate that there was no response at the highest intensities available. (From Zeki 1983a).

wave light depends upon the relative strength of middle wave antagonistic inputs to centre and surround).

A problem that remains unresolved is how the double opponent cells are built up in the monkey. One possibility is that they are built up from the Type I lateral geniculate cells (see Michael, this volume). Another possibility is that they are built up from single opponent cells in the striate cortex itself. The latter either give an ON response to light of one waveband and are unresponsive to lights of other wavebands or give an ON response to some wavelengths and OFF responses to others. They have no antagonistic surrounds and are plentiful in V1 (Zeki 1983a) but relatively rare in the lateral geniculate nucleus (Type IV of Wiesel and Hubel (1966)). Because they do not have centre-surround antagonism, one might predict that their responses will not correlate with colours. It was nevertheless interesting to test the prediction because these cells have action spectra considerably narrower than the cones and it was thus interesting to see whether one could predict which colour a cell would respond best to from a knowledge of its wavelength sensitivity curve. One might suppose, for example, that a wavelength selective cell responsive to middle wave (green) light only will respond best to green areas. It turns out that when studied in experimental conditions similar to the ones used by Land in his perceptual experiments, such cells will respond to an area of any colour if it reflects a sufficient amount of light of their preferred waveband. (The advantage of using the Land paradigm is that the human observer can determine the colour of the area in the cell's receptive field under varying conditions of illumination and thus ascertain whether the responses of the cells correlate with colours). The cell of Figure 8, for example, a middle-wave sensitive cell, could be made to respond to an area of any colour by adjusting the amounts of long, middle and short wave light reflected from the area in its receptive field. Once the reflected energies which yielded a response were determined for one area, the cell could be made to give an almost identical response to areas of other colour if they, in turn, were made to reflect the same triplet of energies. Long wave sensitive cells behaved similarly, i.e. they either responded or did not respond to an area of any colour depending only on whether there was a sufficient amount of long wave light reflected from the area in their receptive fields. (It is worth noting that these cells responded in an identical way if the area of the multicoloured display in their receptive field was isolated from the rest of the display - a void condition in which the colour of the area correlates well with the wavelength composition of the light reflected from it (see Land 1974 and Zeki 1983a)). What role such cells, which I call wavelength selective or WL cells (rather than colour coded ones - an obvious misnomer) play, apart from possibly providing the input to the double-opponent cells, remains mysterious (but see below).

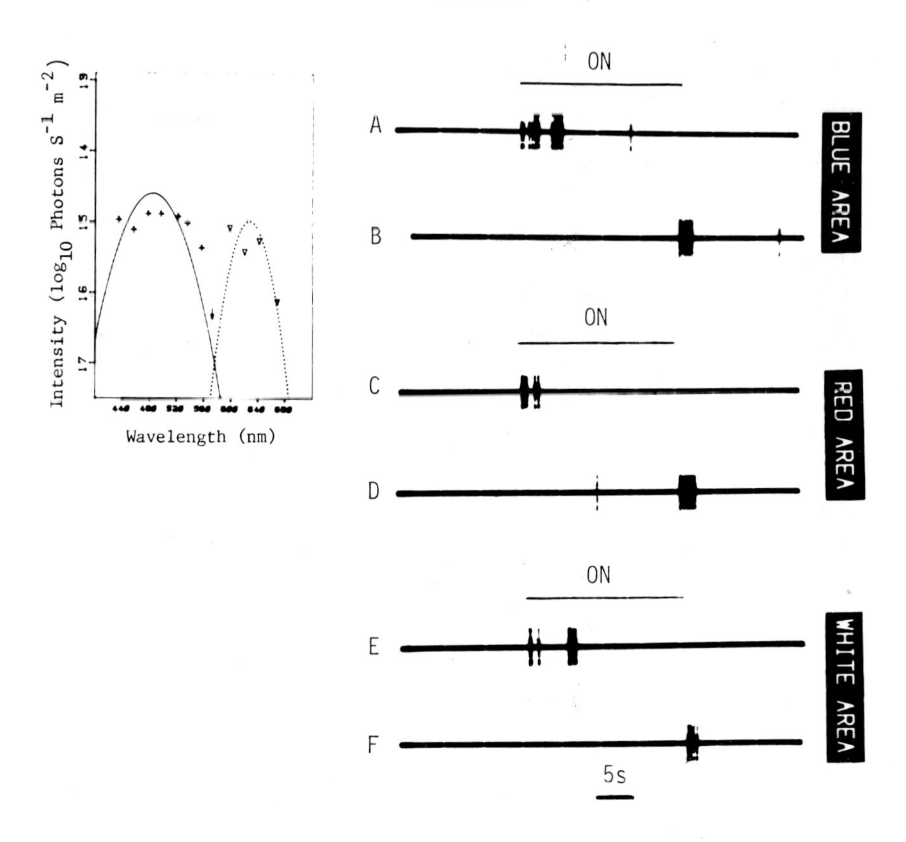

Figure 9: The responses of an opponent input cell in V1 to the blue, red and white areas of a multicoloured display. Inset shows its action spectrum. This cell gave an ON and OFF response to an area of any perceived natural colour. Receptive field size was 0.5° x 0.5°. Stimulation through ipsilateral eye. In A the blue area was made to reflect 40L, 32M and 41S. In B the blue area was made to reflect 69L, 33M and 2S. In C the red area was made to reflect 30L, 20M and 6S. In D the red area was made to reflect 100L, 14M and 5S. In E the white area was made to reflect 43L, 32M and 12S. In F the white area was made to reflect 58L, 33M and 12S. L, M and S refer to long, middle and short wavelength light respectively. All units measured in mW Sr^{-1} m^{-2}. (From Zeki 1983a).

Opponency

Another type of cell in V1 is one which gives explicit ON responses to some wavebands and OFF responses to others from the same part of the receptive field and has no obvious antagonistic surround. These cells, unlike the most common type of wavelength selective cell in the retina or lateral geniculate nucleus (Type I), could thus conceivably account for the Hering perceptual phenomenon of opponent colours in which two opponent colours cannot occupy simultaneously the same position in space (see also Chapters by Hurvich, Jameson and Gouras, this volume). However, it turns out that the reponses of such cells do not obviously correlate with opponent colours.* Figure 9 shows the responses of an opponent input cell in V1 when different areas of a multicoloured display were put in its receptive field and made to reflect varying amounts of long, middle and short wave light. It is notable that the cell could be made to give an ON or an OFF response to an area of any colour, depending only upon the relative amount of different wavelengths reflected from it and independently of its colour. Other cells giving explicit opponent responses behaved similarly and I refer to all such cells as wavelength opponent or WLO cells (rather than opponent colour cells - another obvious misnomer). The behaviour of these cells seems surprising, for it would be difficult to believe that there is no causal relationship between the opponent wavelength pairings observed in single cells and the phenomenal pairing of hues. But such results, in showing that the responses of these cells cannot be the direct basis of the observed perceptual opponencies, raise the question of what the function of opponency may be, in terms of colour vision. One possibility is that having cells with narrow band, opponent, inputs makes the colour system a good deal more sensitive to changes in wavelength composition and hence aid in signalling borders across which there are even small changes in the dominant wavelength reflected. Another possibility is that the fidelity of signalling the presence of a wavelength against a background noise of other wavelengths might be increased. A cell which receives excitation only will signal the presence of a preferred stimulus by an increase over its spontaneous firing rate. If inhibition is added to its repertoire, the discrimination with respect to the presence or absence of a favoured stimulus is enhanced.

* It is worth noting that, just as the colour of a surface bears no simple and obvious relationship to the wavelength composition of the light reflected from it alone, but depends as well on the wavelength composition of the light reflected from surrounding areas, so the colour of the after image produced by viewing an area is dependent upon the wavelength composition of the light reflected from that area as well as surrounding areas. For example, the after image produced by viewing a green area which forms part of a multicoloured display and reflects 30, 60, & 20 mW. Sr^{-1} m^{-2} is red. But the after image produced by viewing the same area when it reflects 60, 30 and 10 mW. Sr^{-1} m^{-2} is also red, even though the amount of long-wave (red) light reflected is now considerably more than middle or short-wave light (see Zeki 1983a and b).

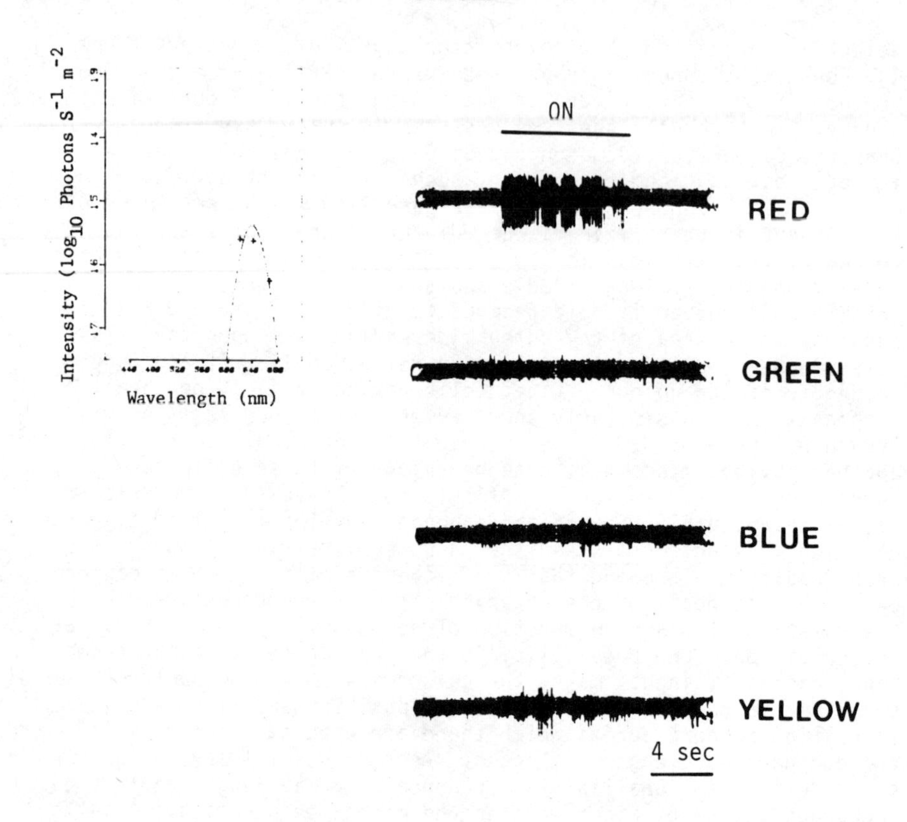

Figure 10: The responses of a long wave selective, colour coded cell in V4 to areas of different colour. Its action spectrum is shown in the inset. Each area when put in the cell's receptive field, was made to reflect the 69, 33 and 7 mW Sr^{-1} m^{-2} of long, middle and short wavelength light respectively. The receptive field of this binocularly driven cell was 3½° x 5° and situated within the central 7°. It was stimulated through the ipsilateral eye. (From Zeki 1983a).

The reaction of colour coded cells

The presence of so many wavelength selective cells whose responses do not correlate with colours was surprising, partly because it is counterintuitive (we see long wave light as red - it therefore becomes surprising to find that a long wave selective cell will respond to an area of any colour which is reflecting a sufficient amount of long wave light) and partly because it runs counter to a tradition that equates wavelength with colour in a rather simple and straightforward manner. It therefore comes as a relief to find that there are cells, in V4, whose responses do correlate with colours. For example, the cell of Figure 10 which, when tested for its wavelength preference, reacted to long wave light only, also responded to the red area of a multicoloured display only and did not respond to areas of other colour even though, when put in the cell's receptive field, each was made to reflect the same triplet of energies as the red area to which the cell responded (Zeki 1980b, 1983a). I refer to such cells as colour coded cells, to distinguish them from the wavelength selective cells described earlier. The difference in the response of these cells and the double opponent cells is almost certainly due to the extensive and complex surrounds of the colour coded cells (Zeki 1980, 1983a Tanaka et al. 1984). That the surround is instrumental in generating the properties of these cells is implicit in their responses (see Zeki 1983a), but it seems certain that, for these cells, the surrounds express themselves through the centre and our attempts to 'dissect out' the surrounds, in the way that one can for the cells of the retina or the lateral geniculate nucleus, have not been successful.

A hierarchy in the colour pathways

One can thus see that, within the segregated cortical colour pathways leading to V4, a hierarchy is in operation. The wavelength selective WL and WLO cells are simpler in their response properties to the wavelength differencing double opponent cells and the latter simpler in their responses compared to the colour coded cells of V4. The hierarchy in responses also correlates with a hierarchy of areas, since colour coded cells have not been found in V1. Where colour coded cells are generated is not known. One possibility is that they are elaborated in V4 itself, which contains both wavelength selective and colour coded cells (Zeki 1983a). Another possibility is that they are partly elaborated in V2. The role of V2 in this respect is still mysterious but it would be difficult to suppose that the information is relayed passively through it.

It is obviously important to relate this hierarchy within a segregated pathway to retinex theory and retinex algorithms. In terms of hierarchies, retinex computations involve, first, the determination of local differences in reflectance for a given waveband, followed next by a global operation in which the reflectance of an area for a given waveband with respect to the average reflectance of the surround for the same waveband is determined and, finally, a colour generation stage in which the results of the global operation for each waveband are compared and the colour thus generated. It is possible that the first stage corresponds roughly to the initial lightness generating operations performed by the double opponent cells but what neurological operations, and where in the cortex, the remaining two operations correspond to remains unresolved. It is possible that the colour generating system is within the V4 complex. There are three main reasons for thinking so. Firstly, the presence of colour coded cells in V4 and their absence in V1 points to V4 (or to V2) as likely candidates. Secondly, psychophysical experiments using a patient with a split corpus callosum (Land et al. 1983) have shown that the corpus callosum is necessary for the colour generating system to integrate information from one visual hemifield for the determination of the colour of a target 3.7° from the midline in the other hemifield. In the monkey, callosal connections at the V1-V2 boundary are restricted very much to the midline and hence would not be involved in integrating information from regions as far as 3.7° into the contralateral hemifield (Whitteridge 1965; Zeki 1970). The first cortical area in which there are wavelength and colour coded cells and in which the callosal connections unite cells whose fields extend well beyond 5° is the V4 complex (Zeki 1970; 1977). Assuming that the organization of the visual areas in the monkey is similar to that in man, this makes of the V4 complex a good candidate for the colour generating system. Finally, the internal wiring within V1 and V2 is limited tangentially (Fisken et al. 1975 and Zeki, unpublished results) which implies that the long range interactions necessary for the colour generating system are absent within these two areas. By contrast, there are powerful tangential connections within the V4 complex, enabling a cell located in one part of the V4 complex to be informed of the activity of another cell, removed by several millimetres (more than 4) from it and whose receptive field is several degrees away. This is not to suggest that the neurological correlate of the lightness generating system does not also operate, at least in part, within the V4 complex. The global lightness operations also require long-range interactions and the V4 complex would be suited for such a role.

In general, then, one can see that retinex theory, more than the classical theories of colour vision, can provide a framework for cortical experiments in colour vision. But this does not mean that the mode of operation of the nervous system in generating colours is identical, or even similar, to that postulated by retinex theory. Indeed the opposite is the case for retinex theory supposes that there are three channels, identical to the cone channels, which are maintained separate from the retina to the comparison site somewhere in the central nervous system. The physiological evidence, by contrast, is unequivocal in showing that the signals are mixed at the post-receptoral level, with the input from one set of cones being opposed by an input of opposite polarity from another (Zrenner 1984). In retinex theory, there is no place for opponency and all colours can be generated by comparing lightnesses generated by each channel. The physiological evidence, by contrast, shows that opponency is an ubiquitous phenomenon of the colour pathways, even if the recent evidence that is reviewed above raises the question of the precise function of such an opponence system. Finally, retinex theory supposes that there are three channels that input to the comparison stage. If one were to judge by the physiological evidence on the peak sensitivities of the double opponent cells (which, to repeat, probably represent an initial stage in the lightness generating system) one could make an argument for the presence of multiple channels that input to the comparison system in the cortex because the peak sensitivities of such cells, unlike the absorption spectra of the cones, are widely distributed throughout the visible spectrum (McNichol et al. 1964; Bowmaker et al. 1980; Michael 1978). Hence there are differences between the mechanisms postulated by retinex theory and those used by the nervous system to generate colours. But the end result of both is the same. In other words, whereas the details of the implementation may differ, both types of operation lead to the generation of colour and, moreover, to colours which change little, if at all, with profound changes in the wavelength composition reflected from a single surface.

Conclusion:

The demonstration of functional specialization within the prestriate visual cortex naturally raises the question of what useful purpose it serves and why different kinds of visual information cannot be processed in the same area. Retinex algorithms provide some hint (and it is interesting to note that retinex theory came to be seriously considered by physiologists only after the demonstration of functional specialization). To determine the colour of a surface, the nervous system must compare, simultaneously, the wavelength composition of the light reflected from that surface in relation to the wavelength composition of the light reflected from all other surfaces in the field of view. In this scheme, the precise disposition of the surface in relation to one another is not critical. To determine motion within the field of view, the nervous system has to determine the sequential activation, in time, of different parts of the field of view. Hence, among other features, the time requirements for determining colour and motion are sufficiently different for the nervous system to use different programs, pathways and areas for each. Again, the precise relations of adjacent points and surfaces to one another are critical in the constructions of forms, but not in the construction of colour and this, too, presumably requires a different kind of algorithm. And hence, presumably, the need for different visual areas to process different kinds of visual information.

The details of the operations that the nervous system performs to construct different attributes of the visual scene and then to combine them to give our unitary perception of the visual world are not known. Major areas remain to be explored. But recent advances in cerebral physiology and colour theory are showing that, in constructing an image of the visual world, the brain has to solve a far more intricate and subtle problem than we had imagined. It would be surprising if this insight does not lead to profound changes in the question that we ask experimentally of the nervous system.

The author's work was supported by grants from the Science and Engineering Research Council.

REFERENCES

Andersen, V.O., Guld, C. and Sjö, O. (1983). Colour processing in prestriate cortex of vervet monkey. In: Mollon J.D., Taylor, S.K. (eds) Colour Vision. Academic Press, London.

Baizer, J.S. (1982). Receptive field properties of V3 neurons in monkey. Invest. Opthal. 23, 87-95.

Campbell, A.W. (1905). Histological studies on the localization of cerebral function. Cambridge University Press, Cambridge.

Clare, M.H. and Bishop, G.H. (1954). Responses from an association area secondarily activated from optic cortex. J. Neurophysiol. 17, 271-277.

Cragg, B.G. (1969). The topography of the afferent projections in circumstriate visual cortex of the monkey studied by the Nauta method. Vision Res. 9, 733-747.

Damassio, A., Yamada, T., Damasio, H. Corbett, J. and McKee, J. (1980). Central achromatopsia: Behavioural, anatomic and physiological aspects. Neurology 30, 1064-1071.

Daw, N.W. (1967). Neurophysiology of colour vision. Physiol. Rev. 53, 571-611.

DeValois, R.L. (1972). Processing of intensity and wavelength information in the visual system. Invest. Opthal. 11, 417-427.

Fisken, R.A., Garey, L.J. and Powell, T.P.S. (1975). The intrinsic assoication and commissural connections of area 17 of the visual cortex. Phil. Trans. R. Soc. Lond. B. 272, 487-536.

Fries, W. and Zeki, S.M. (1979). Effect of bilateral prestriate cortex (V4) lesions on wavelength discrimination in monkeys. Pflugers Arch 382, R. 46.

Fries, W. and Zeki, S. (1983). The laminar origin of the cortical inputs to the fourth visual complex of the macaque monkey cortex. J. Physiol. Lond. 340, 51P.

Gattass, R. and Gross C.G. (1981). Visual topography of striate projection zone (MT) in posterior superior temporal sulcus of the macaque. J. Neurophysiol. 46, 621-638.

Hochberg, J.E. (1964). Perception. Prentice-Hall, Englewood Ciffs, New Jersey.

Horn, B.K.P. (1974). Determining lightnesses from an image. Computer Graphics and Image Processing. 3, 277-299.

Horton, J.C. and Hubel, D.H. (1981). Regular patchy distribution of cytochrome oxidase staining in primary visual cortex of macaque monkey. Nature, London 292, 762-764.

Hubel, D.H. and Wiesel, T.N. (1965). Receptive fields and functional architecture in two non-striate visual areas (18 and 19) of the cat. J. Neurophysiol. 28, 229-289.

Hubel, D.H. and Wiesel, T.N. (1969). Visual area of the lateral suprasylvian gyrus (Clare-Bishop area) of the cat. J. Physiol. Lond. 202, 251-260.

Hubel, D.H. and Wiesel, T.N. (1974). Sequence regularity and geometry of orientation columns in the monkey striate cortex. J. Comp. Neurol. 158, 267-294.

Hubel, D.H. and Wiesel, T.N. (1977). Ferrier Lecture: Functional architecture of macaque monkey visual cortex. Proc. Roy. Soc. Lond. B, 198, 1-59.

Land, E.H. (1974). The retinex theory of colour vision. Proc. of the Royal Inst. of Gt. Britain, 47, 23.

Land, E.H. (1983). Color vision and the natural image. III. Recent advances in retinex theory and some implications for cortical computations. Proc. Natl. Acad. Sci. USA. 80, 5163-5169.

Land, E.H., Hubel, D.H., Livingstone, M.S., Perry, S.H. and Burnes, M.M. (1983). Colour generating interactions across the corpus callosum. Nature Lond. 303, 616-618.

Livingstone, M.S. and Hubel, D.H. (1982). Thalamic inputs to cytochrome oxidase-rich regions in monkey visual cortex. Proc. Natl. Acad. Sci. USA. 79, 6098-6101.

Livingstone, M.S. and Hubel, D.H. (1984). Anatomy and physiology of a color system in the primate visual cortex. J. Neurosci. 4, 309-356.

Lund, J.S., Lund, R.D., Hendrickson, A.E., Bunt, A.H. and Fuchs, A.F. (1975). The origin of efferent pathways from the primary visual cortex, area 17, of the macaque monkey as shown by retrograde transport of horse radish peroxidase. J. Comp. Neurol. 164, 287-304.

Michael, C.R. (1978). Color vision mechanisms in monkey striate cortex: dual opponent cells with concentric receptive fields. J. Neurophysiol. 41, 572-588.

Michael, C.R. (1981). Columnar organisation of colour cells in monkey striate cortex. J. Neurophysiol. 46, 587-604.

Meadows, J.C. (1974). Disturbed perception of colours associated with localised cerebral lesions. Brain, 97, 615-632.

Morgan, A.A. (1966). Chromatic adaptation in the macaque. J. Comp. Physiol. Psychol. 62, 76-83.

Otsuka, R. and Hassler, R. (1962). Über aufbau und fleiderung der corticalen sehsphäre bei der katze. Arch. Psychiat. Nervenkr. 203, 212-234.

Pearlman, A., Birch, J. and Meadows, J.C. (1979). Cerebral color blindness: An acquired defect in hue discrimination. Ann. Neurol. 5, 253-261.

Poggio, G.F., Baker, F.H. Mansfield, R.J.W., Sillito, A. and Grigg, P. (1975). Spatial and chromatic properties of neurons subserving foveal and parafoveal vision in rhesus monkey. Brain Res. 100 , 25-59.

Shipp, S. and Zeki, S. (1985). The segregation of pathways leading to V5 in macaque monkey visual cortex. Nature Lond. In press.

Tanaka, M., Creutzfeldt, O. Werer, H. and Lee, B. (1984). Visual responses of single units in the prelunate gyrus of the awake monkey. Neuroscience Letters Supp. 18, 571.

Ungerleider, L.G. and Mishkin, M. (1979). The striate projection zone in the superior temporal sulcus of Macaca mulatta: location and topographic organisation. J. Comp. Neurol. 188, 347-366.

Van Essen, D.C., Maunsell, J.H.R. and Bixby, J.L. (1981). The middle temporal visual area in the macaque: myeloarchitecture, connections, functional properties and topographic organization. J. Comp. Neurol. 199, 293-326.

Von Economo, C. and Koskinas, G.N. (1925). Cytoarchitecktonik der Grosshirninde des erwachsenen Menschen. Springer, Berlin.

Whitteridge, D. (1965). Area 18 and the vertical meridian of vision. In: Ettlinger, E.G. (ed). Functions of the Corpus Callosum. Churchill, Lond.

Wiesel, T.N. and Hubel, D.H. (1966). Spatial and chromatic interactions in the lateral geniculate body of the rhesus monkey. J. Neurophysiol. 29, 1115-1156.

Wild, H., Butler, S.R., Carden, B. and Kulikowski, J.J. (1984). Primate cortical area V4 is important for colour constancy but not wavelength discrimination. Nature, Lond. In press.

Wong-Riley, M.T.T. (1979). Changes in the visual system of monocularly sutured or enucleated cats demonstrable with cytochrome oxidase histochemistry. Brain Res. 171, 11-28.

Wright, M.J. (1969). Visual receptive fields of cells in a cortical area remote from the striate cortex of the cat. Nature, Lond. 223, 973-975.

Wurtz, R.H., Mikami, A., Newsome W.T. and Dursteler, M.R. (1984). In: Perspectives of Neuroscience: from molecule to mind. Y. Tsukada (ed), Tokyo, University Press.

Zeki, S.M. (1969). The secondary visual areas of the monkey. Brain Res. 13, 197-226.

Zeki, S.M. (1969). The representation of central visual fields in prestriate cortex of monkeys. Brain Res. 14, 271-291.

Zeki, S.M. (1970). Interhemispheric connections of prestriate cortex of monkey. Brain Res. 19, 63-75.

Zeki, S.M. (1971). Cortical projections from two prestriate areas in the monkey. Brain Res. 34, 19-35.

Zeki, S.M. (1973). Colour coding in rhesus monkey prestriate cortex. Brain Res. 53, 422-427.

Zeki, S.M. (1974). Functional organisation of a visual area in the posterior bank of the superior temporal sulcus of the rhesus monkey. J. Physiol. Lond. 236, 549-573.

Zeki, S.M. (1975). The functional organisation of projections from striate to prestriate visual cortex in the rhesus monkey. Cold Spring Harbor Symp. Quant. Biol. 40, 591-600.

Zeki, S.M. (1976). The projections to the superior temporal sulcus from areas 17 and 18 in the rhesus monkey. Proc. Roy. Soc. Lond. B. 193, 199-207.

Zeki, S.M. (1977). Colour coding in the superior temporal sulcus of rhesus monkey visual cortex. Proc. Roy. Soc. Lond. B. 195, 517-523.

Zeki, S. (1978). Uniformity and diversity of structure and function in rhesus monkey prestriate visual cortex. J. Physiol. Lond. 277, 273-290.

Zeki, S. (1978). Functional specialisation in the visual cortex of the rhesus monkey. Nature, Lond. 274, 423-428.

Zeki, S. (1978a). The third visual complex of rhesus monkey prestriate cortex. J. Physiol. Lond. 277, 245-272.

Zeki, S. (1980). The representation of colours in the cerebral cortex of the monkey. Nature, Lond. 284, 412-418.

Zeki, S. (1980b). The consequence of varying background colours in stimulating cortical colour-coded cells. J. Physiol. Lond. P. 305, 71-72.

Zeki, S. (1981). The mapping of visual functions in the cerebral cortex. Proceedings of the Third Tanaguchi Foundation Symposium. R. Norgren and M. Sato (eds). John Wiley and Sons, New York.

Zeki, S. (1983). Does the colour of the after-image depend upon wavelength composition? J. Physiol. Lond. (P) 338, 36.

Zeki, S. (1983a). Colour coding in the cerebral cortex: The reaction of cells in monkey visual cortex to wavelengths and colours. Neurosci. 9, No. 4, 741-765.

Zeki, S. (1983b). Colour coding in the cerebral cortex: The responses of wavelength-selective and colour-coded cells in monkey visual cortex to changes in wavelength composition. Neurosci. 9, No. 4, 767-781.

Zeki, S. (1983). The distribution of wavelength and orientation selective cells in different areas of monkey visual cortex. Proc. Roy. Soc. Lond. B. 217, 449-490.

Zeki, S. and Shipp, S. (1985). Visual physiology and colour theory. Nature, Lond. (In press).

Zihl, J., Von Cramon, D, and Mai, N. (1983). Selective disturbance of movement vision after bialateral brain damage. Brain 106, 313-340.

Zrenner, E. (1983). Neurophysiological aspects of colour vision in primates. Springer, Berlin.

On Lightness Computation in Mondrian World

A. Blake

Department of Computer Science, University of Edinburgh, Edinburgh, UK

ABSTRACT

The retinex theory of lightness computation explains how for a "Mondrian World" image, consisting of a number of patches each of uniform reflectance, the reflectances can be computed from an image of that object. The computation can be realised as a parallel process performed by successive layers of cooperating computational cells, arranged on hexagonal grids. The layers will, in practice, be of finite extent and it is critical that cells on the array boundary behave correctly. The process has been successfully implemented on a digital computer.

1. RETINEX THEORY

The Retinex theory of lightness perception is described in Land and McCann (1971) and in Land (1983). It explains how it is possible, in the restricted world of "Mondrian" images, to compute lightness – the psychophysical correlate of surface reflectance – even under highly non-uniform illumination. The theory was motivated by the notable ability of human vision to perceive surface reflectance in apparently adverse illumination conditions: a uniform, white piece of paper continues to look white when moved into shadow, and still appears to be of uniform colour even under a steep illumination gradient (for example, from a nearby point source). Moreover, the theory explains some but not all simultaneous contrast illusions (Marr 1982).

A Mondrian image is defined to consist entirely of patches of uniform reflectance, separated by step edges. In such a restricted world, reflectance variations are not confounded with illumination variations: the reflectance function is piecewise constant – all its variations are assumed to be step changes – whereas the illumination function is assumed everywhere to have finite gradient. Image intensity is the product of reflectance and illumination so that step changes in intensity indicate step changes in reflectance and any gradual change in intensity corresponds to a variation in illumination. Thus, in principle, the reflectance and illumination functions can be recovered from the image intensity.

Land and McCann propose to perform the recovery in the following way. First colour is treated by identical but independent retinex processes in three colour channels, so that each retinex is effectively a monochrome process. Now, given one of the three intensity functions, step changes in intensity are detected. By measuring the *ratio* of intensities on either side of a step, all information about reflectance is extracted. Within boundaries delineated by step changes, reflectance is constant and the ratio of reflectances in adjacent patches is given by the intensity ratio across their common boundary. The reflectance function can be reconstructed explicitly by first choosing one patch arbitrarily to have reflectance value 1, then the reflectance of any other patch is determined by following a path back to the 1st patch and calculating the product of intensity ratios across all boundary crossed. The result is independent of the path chosen, for Mondrian images. In this way the reflectance is determined up to a multiplicative constant; Land and McCann suggest that, in a monochrome image, the reflectance function be normalised by finding the patch of highest reflectance in the image, and calling it white. More recently however, Land (1983) has demonstrated an alternative method that involves, for each patch, computing reflectance relative to a number of other patches. For a given waveband, the average of those relative reflectances then forms an absolute designator for that patch and waveband.

Horn (1974) developed a cooperative algorithm for lightness computation based on the retinex. Given an intensity function

$$p'(x,y) = r'(x,y)s'(x,y)$$

where r',s' are reflectance and illumination functions, it is convenient to work with their logarithms ($p = log(p')$ etc.) so that

$$p(x,y) = r(x,y)+s(x,y) \tag{1}$$

– r and s are linearly separable. In other words the reflectance and illumination components, which were combined multiplicatively, are now combined additively. As a result, the effect of applying some gradient operator (e.g. L below) to the combined signal p is equivalent simply to applying that operator to r and s individually. Horn's algorithm is simply to compute a lightness image

$$l = L^{-1}TLp \tag{2}$$

where L,L^{-1} are respectively the laplacian operator ∇^2, in two dimensions, and its inverse. The operator T simply thresholds: given a function f and a fixed threshold λ

$$(Tf)(x,y) = f(x,y) \text{ if } |f(x,y)|>\lambda, \text{ 0 otherwise.}$$

The effect of T in (2) is to preserve sharp changes in p but to eliminate slow ones; this supresses the s component of p, leaving the r component In a discrete image, in which intensity values are held on a square or hexagonal tessalation of cells (*pixels*), the laplacian L is simply a local convolution (fig 1a). The inverse laplacian L^{-1} can be applied by iterative convolution, in which the value at each pixel is repeatedly recomputed to be: the value of the *input* (i.e. the signal whose inverse Laplacian is being computed) plus the average of the *output* values at neighbouring pixels (fig 1b).

In this paper we point out that Horn's algorithm is not entirely equivalent to Land and McCann's retinex computation, nor will it succeed in all cases to extract the reflectance component of the intensity distribution of a Mondrian image. The difficulty is in imposing correct constraints on the reflectance within each patch: reflectance within a patch is known to be uniform, whereas in Horn's scheme it may, for instance, have any linear form – in fact it is constrained only to be a harmonic function.

We show in this paper how the stronger constraint can be used to correctly recover reflectance in Mondrian world. It transpires that the new scheme also uses thresholding and computation of an inverse laplacian. Therefore the arguments of Marr (1974) for biological feasiblility of Horn's scheme should still apply.

Briefly, the computation is:

1. Compute the gradient $F = \nabla p$

2. Threshold: $E = T(F)$ to remove the component of the gradient that is due to varying illumination, leaving only the gradient of reflectance. In contrast, Horn's method involved thresholding the output of a Laplacian rather than a gradient operator.

3. Recover the reflectance from its gradient. This can be done by a parallel, iterative method that involves computing the divergence of the gradient (i.e. differentiating it again) and then an inverse Laplacian, under special boundary conditions. The entire computation can be performed in a highly parallel manner, using several layers of computational cells.

Our claim, in the succeeding sections, is that this sequence correctly performs the retinex computation.

This computation is represented in figure 5 (to be found towards the end of this paper) as a sequence of layers of computational cells; this figure and the discussion in section 3.2 should prove helpful for the non-mathematical reader. Cells within layers are arranged on a hexagonal grid and it can be seen that feedback is employed in the final layer to perform the inverse laplacian. The signal in this final layer can be visualised as propagating from cell to cell. Feedback paths connect cells to their neighbours and propagation occurs as signals radiate outward from each cell, via these paths. In theory, the distance over which such signals can travel is unlimited except by the boundaries of the array of cells. In fact this is why boundary conditions are so important – the effect of erroneous treatment of the array boundary propagates

Figure 1

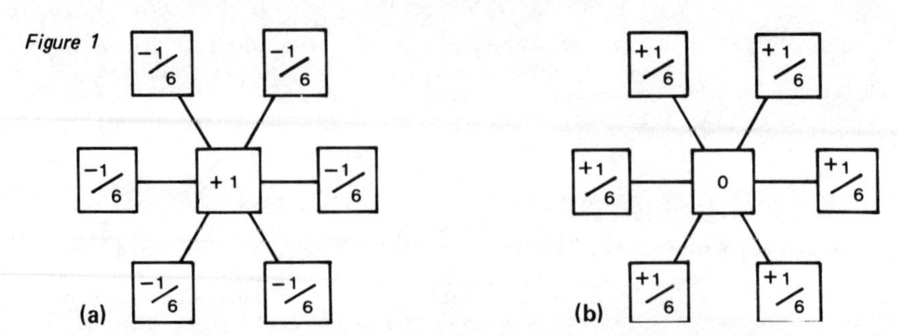

1. The laplacian operator is represented discretely as a local convolution (a) and can be inverted by an iterative process - each iteration consists of applying the convolution in (b) to the output array and adding in the input array.

Figure 2

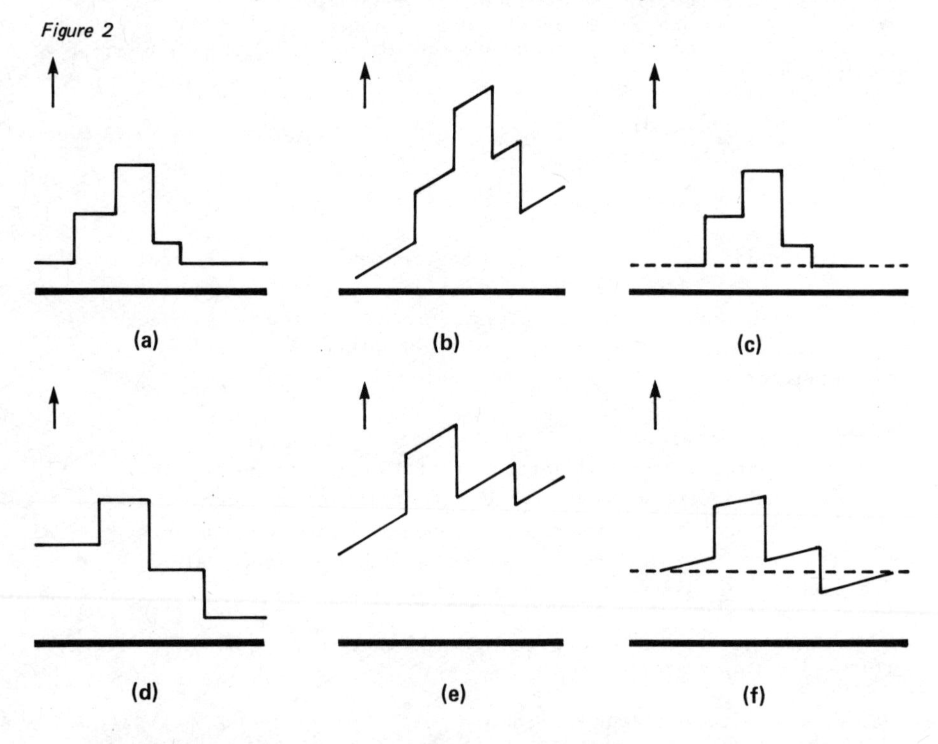

2. Lightness computation (in 1 dimension) by inverting the laplacian under the somewhat arbitrary condition that lightness is zero on boundaries. (In the 1-D case the boundary just consists of two endpoints). The reflectance function (a) under an illumination gradient produces image intensity (b) from which lightness (c) is computed - in this case a correct reconstruction of the reflectance. But the same process applied to (d,e,f) does not reconstruct the reflectance correctly. This is because the boundary conditions are inappropriate.

throughout the rest of the array. In practice, both time available and precision of computation may limit this distance. The computation has been implemented on a digital computer, and the results can be seen in figure 6.

2. THE CONTINUOUS PROBLEM

The purpose of this section is to explain in greater detail why Horn's lightness computation cannot deal with all Mondrian world images and to propose a method as outlined in the introduction, which correctly infers lightness for Mondrian world images.

2.1. Harmonic constraints are too weak

The assumption made in Mondrian world is that the reflectance function is piecewise constant – it consists of regions of constant value, in which $\nabla r=0$, bounded by step changes. However Horn's reconstruction formula (2) imposes only the harmonic constraint on l, within regions, instead of the stronger uniformity constraint. Under the harmonic constraint the inversion formula (2) is not uniquely specified: rewriting (2) as

$$L\,l = TLp \tag{3}$$

it is clear that if, for some g, $l=g$ satisifies (3) in some domain $D \subset R^2$ then for any "harmonic" function f (one that satisfies $\nabla^2 f=0$,) $l=g+f$ also satisifies (3). For instance, any linear function $f(x,y)=ax+by+c$ is harmonic. Classically, the solution of a Poisson equation such as (3) becomes uniquely determined when boundary conditions are specified for l – for example that $l=0$ or (for uniqueness up to a constant) that $\mathbf{n}.\nabla l=0$ on C (the boundary of D), where $\mathbf{n}$ is the normal to C. In the context of lightness computation, it is not immediately clear what boundary conditions on l are appropriate. Suppose that l is always, as required, a correct reconstruction of r, up to an additive constant K, so that

$$l = r+K.$$

The condition

$$l = 0 \text{ on } C$$

for example would imply that r is also constant on C – that the image has a border of constant reflectance – and this is not generally true in Mondrian world. Horn (1974) explicitly assumes that the border does have constant reflectance and therefore his reconstruction algorithm succeeds only when this assumption is valid. If such a boundary condition is used with images which do not have a constant reflectance border then the reconstruction process fails. This is illustrated for one-dimensional images in fig 2.

The problem lies in the application in (2) of the laplacian which, because it is a second order operator, throws away some gradient information. However, we argue that it is admissible to use a gradient operator in place of the laplacian, thus retaining all information about p, up to an additive constant.

2.2. Gradient operator

It is natural to require that the lightness computation should be specified in a way that is coordinate frame invariant: under transformation of the coordinate frame in which the computation is performed the resulting lightness function l should not vary – it is a scalar field. This behaviour can be ensured by requiring all operators used in the computation to be coordinate frame invariant. In particular the operator used initially to extract gradient information from p must be a tensor operator; the operators ∇ and ∇^2 are both examples of tensor operators, of rank 1 (vector) and 0 (scalar) respectively. In Horn's case, $\nabla^2 p$ was used because it is isotropic (as any scalar is isotropic – because it is coordinate-frame invariant). Here we instead use ∇p and the corresponding scalar is $|\nabla p|^2$ – the quantity used in thresholding, as described below.

We use ∇ in place of $L \equiv \nabla^2$, replacing the equation (3) by

$$\nabla l = E \text{ where } E = T\nabla p. \tag{4}$$

Figure 3

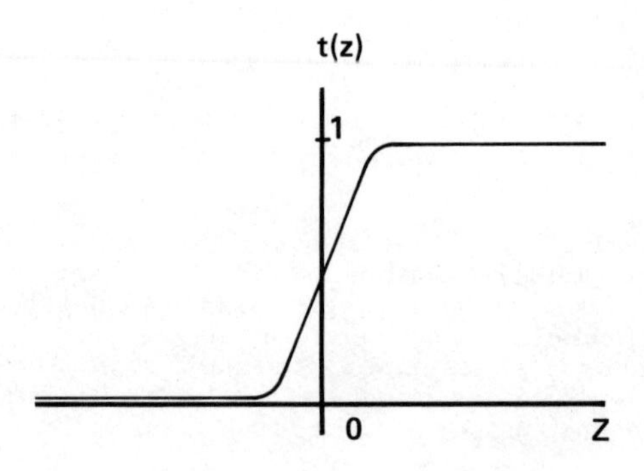

3. A smooth approximation to a step function is used for thresholding the gradient of image intensity.

The thresholding operator T, applied now to a vector field, is defined by:

$$T(A) = t(|A|^2-\lambda^2)A$$

where t is a smooth approximation to the function

$$t(z) = 1 \text{ if } z>0, 0 \text{ otherwise, as in fig 3.}$$

In equation (4), ∇p is a vector field, hence $T\nabla p$ is a vector field in which small gradients of p have been suppressed, but step changes are retained.

It can be shown that a solution for l in (4) does exist and it is clear that, since ∇l is completely specified, l is unique up to a constant – exactly as in Land and McCann's lightness computation.

2.3. Solving the gradient equation

The equation (4) for l can be solved by first assigning a value arbitrarily to l at some point (x_0,y_0) and then integrating directly over paths P from (x_0,y_0) to each point (x,y):

$$l(x,y) = \int_P E.\mathrm{d}r \tag{5}$$

where, provided that the field E is "conservative" (Clemmow, 1973), the result of the integral is dependent on the positions of the endpoints but independent of the path taken between them. Essentially, (5) is a formal statement of Land and McCann's method. Computationally, when the lightness function is represented by a discrete array l_{ij}, it can be reconstructed by propagation, in a single pass, from (x_0,y_0) outwards. However, provided the field E is conservative, it can alternatively be computed using a divergence (differentiation) operator followed by an inverse laplacian, under special boundary conditions – as the solution of the following equation:

$$\nabla^2 l = \nabla.E \text{ in } D, \text{ with Neumann condition } \boldsymbol{n}.\nabla l = \boldsymbol{n}.E \text{ on } C. \tag{6}$$

This equation is known to have a unique solution (up to a constant) (Clemmow 1973).

This method has the advantages that:

- The machinery to compute this inverse was argued by Marr (1974) to be biologically feasible. The whole computation can be performed in a local parallel manner.

- The computation is robust in the sense that, even if condition that the field E is conservative is not obeyed exactly, an optimal approximation for l is found. This arises when the lightness function is represented approximately, by a discrete array (see section 3).

2.4. Robustness

Equation (6) is a robust statement of the lightness computation problem in this sense: even if the condition that E is conservative holds only approximately, the solution for l can be shown to be optimal in that it minimises the quantity

$$F = \int_D (\nabla l - E)^2. \tag{7}$$

The influence on the lightness $l(x,y)$ of errors in the computation of E (these are likely to arise from rounding in limited precision arithmetic and from the approximation involved in discretisation of space) are thus "spread out" in an optimal fashion. However, in reconstruction by direct integration (5), such errors would make the computed lightness function dependent on the path of integration. The lightness assigned to a given region would depend on the path taken from the starting point (x_0,y_0) to the region - different paths could yield mutually inconsistent values of lightness for that region.

That completes our discussion of the computation of lightness as a function in euclidean space. The next section translates the results we have just seen into the discrete domain, in which scalar fields are represented as arrays.

3. DISCRETE COMPUTATION

3.1. The problem

In order to be able to compute lightness for a sampled image, the problem in the previous chapter must be expressed in terms of arrays rather than scalar fields. We will take the problem statement in variational form (7) and replace the minimisation of the integral by a corresponding minimisation of a sum over array elements. Using linear, equilateral triangular elements (Strang and Fix, 1973) we will obtain a discrete version of equation (6), in which the laplacian operator appears as a convolution in a hexagonally connected array, as in Horn (1974) and Marr (1974).

First we introduce notation for the discrete fields. The tessellation Z is a set of of equilateral triangular elements, as shown in fig 4a, with vertices labelled by pairs (i,j). The triangles themselves are labelled by two sets of vertices:

$$U = \{(i,j) : \text{the triangle } above \ (i,j) \text{ is in } Z\}$$

$$D = \{(i,j) : \text{the triangle } below \ (i,j) \text{ is in } Z\}$$

Let V be the set of all vertices of triangles in Z.

We denote the set of near neighbours (6 of them but fewer on the

Figure 4a

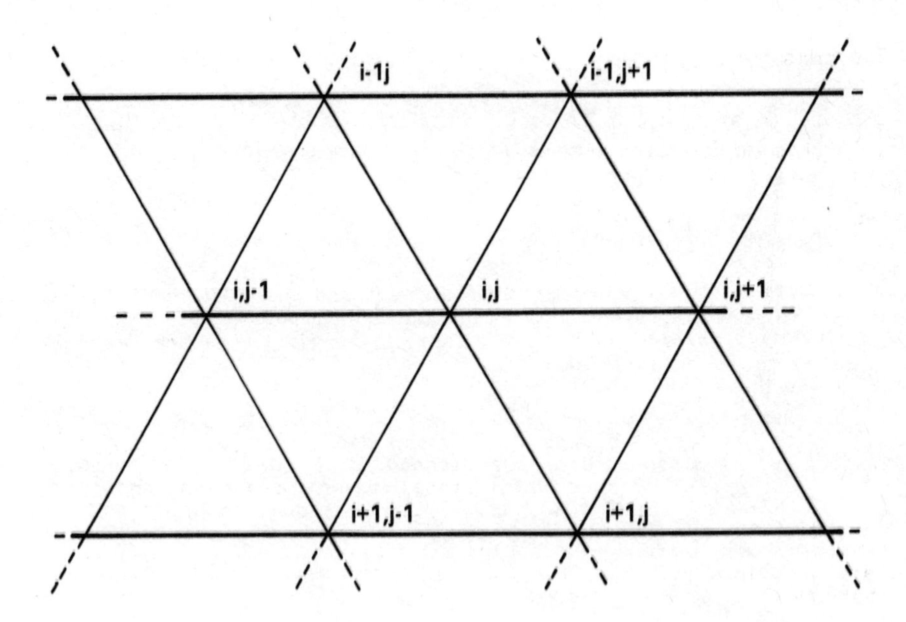

Figure 4b

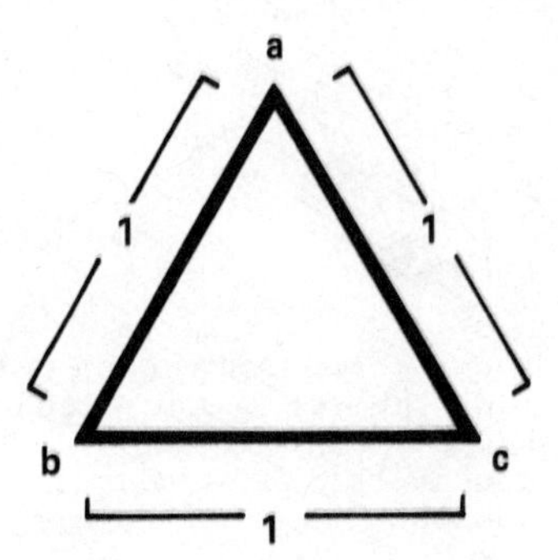

4. The vertices in a tessalation (a) of equilateral triangles are used as coordinates for lightness computation. The vertices are indexed as shown. One triangle (b) of the tessalation is the support of a piecewise linear basis function used in discrete approximation by finite element. The function is linear inside the triangle and zero outside. It has values a,b,c at its vertices.

array boundary) of (i,j) by $N(i,j)$. The quantities represented by arrays are:

- p_{ij} is the log-intensity, assumed to be *sampled* at i,j.

- u_{ij}, d_{ij} are boolean "threshold flags" that indicate whether the gradient in the triangles respectively above and below (i,j) exceeds threshold λ.

- l_{ij} is the reconstructed log-reflectance, also sampled at (i,j).

We assume, in the spirit of finite element analysis, that the fields p,l are adequately represented by the linear interpolants of p_{ij},l_{ij} (the interpolating functions that agree with p_{ij},l_{ij} at triangle vertices and are linear within each triangle).

The energy F (7) can now be expressed in terms of the discrete arrays $u_{i,j}$, $d_{i,j}$, $l_{i,j}$ and $p_{i,j}$. Setting to zero the partial derivatives of F with respect to $l_{i,j}$ then yields linear simultaneous equations for $l_{i,j}$ in terms of $u_{i,j}$, $d_{i,j}$ and $p_{i,j}$. These simultaneous equations are the discrete equivalent of the differential equation (6). They incorporate discrete versions of the divergence and the Laplacian operators, and also the boundary conditions.

3.2. The solution

Having defined the discrete equations, in terms of a discrete laplacian operator, it is necessary to solve them. This can be done in parallel, by iterative convolution in a similar way to Horn (1974). This solution is graphically illustrated in figure 5 as a sequence of layers of processing cells, through which the intensity signal $p_{i,j}$ is passed to obtain the lightness signal $l_{i,j}$. First we illustrate a simplified sequence of layers (fig 5a) for processing a one dimensional signal; then (fig 5b) for a full two-dimensional signal, as considered in this paper. It can be seen that the connections between the hexagonally connected layers in fig 5b are quite complex, with computational cells in layer 5 having 18 inputs each.

The boundary conditions, described in preceding sections, can be implemented quite naturally by the computational mechanism of figure 5b. Cells in layer 2 and 3 are omitted if any of their inputs are absent. Cells in layer 4 however may have less than the full number of inputs (from layers 3 and 4), but then weights on the inputs to that cell must be altered accordingly.

A solution using Gauss-Seidel relaxation (a serial variant of Jacobi relaxation - see below) has been implemented on a serial computer (an Edinburgh University Advanced Personal Machine). Typically, with 64x64 pixel images, about 2000 iterations are used. Results are shown in figure 6.

Finally, some technical points: when the parallel convolutions are also synchronous throughout the tessellation, this process is essentially Jacobi relaxation, in which the nth iteration of the lightness array is given by:

Figure 5a

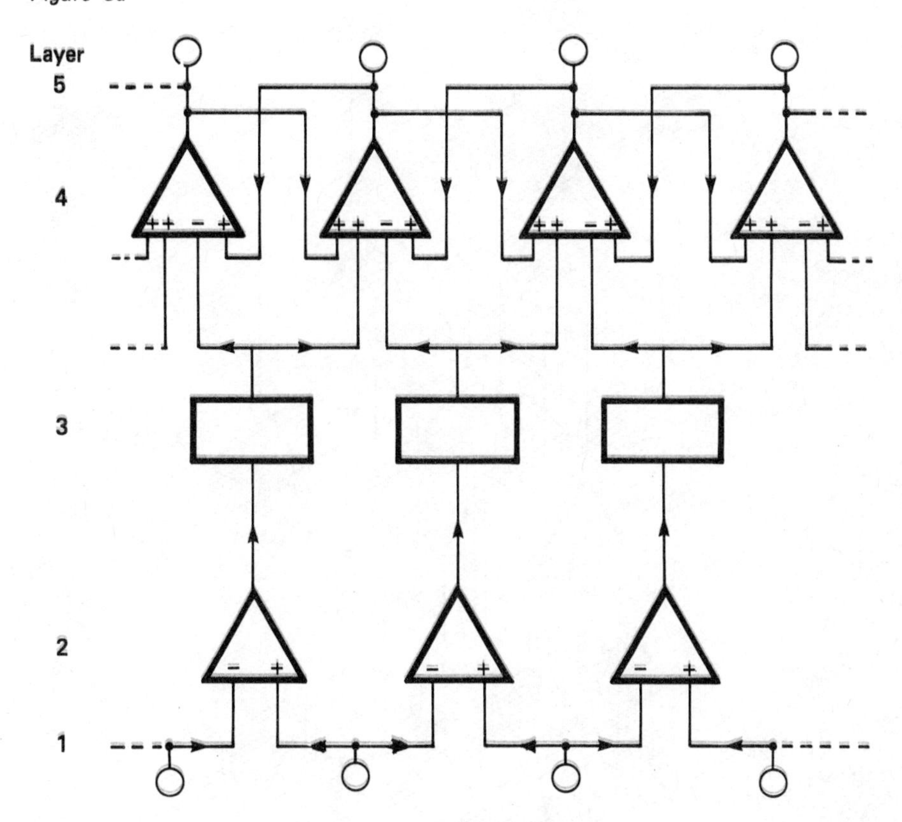

5. **The possible structure of hardware to compute lightness.** For simplicity we first illustrate (a) this for one-dimensional data. Layer 1 is data input, layers 2 & 4 contain cells that output a weighted sum of their inputs - weights may be positive or negative. Layer 2 cells compute the gradient of the input signal. Layer 3 contains thresholding units that transmit the input signal to the output, if the signal magnitude exceeds a certain threshold. Layer 4 performs a dual function: it computes the divergence of the signal in layer 3 (i.e. differentiates it again); then the feedback from other cells in layer 4 sets up a cooperative process to compute an inverse Laplacian. Layer 5 contains the output - i.e. the lightness array. For two-dimensional data (b) connections between cells are considerably more complex, for instance layer 4 cells each have 18 inputs. However the function of each layer is as in (a) except that a layer 3 cell now has 3 inputs instead of 1; it transmits each of its 3 inputs if the sum of the squares of those inputs exceeds threshold.

Figure 5b

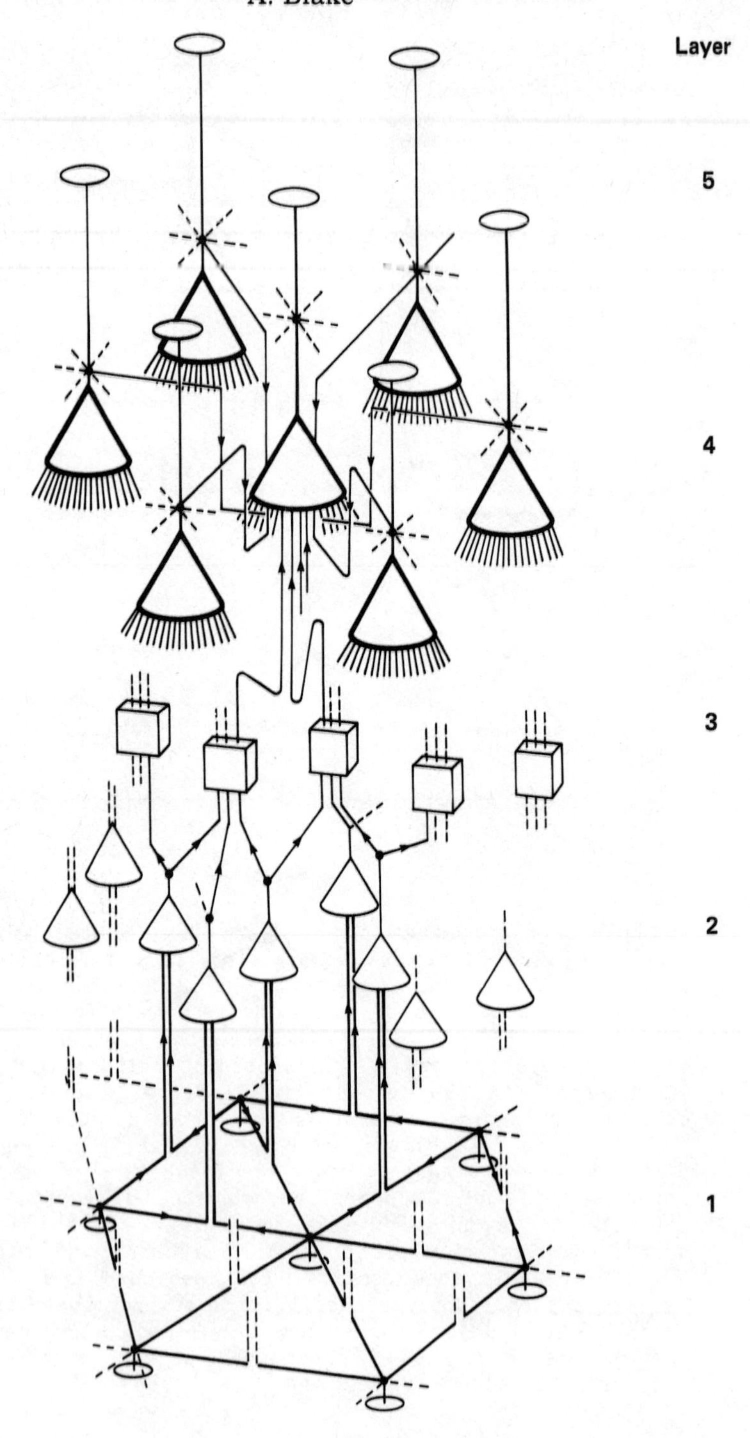

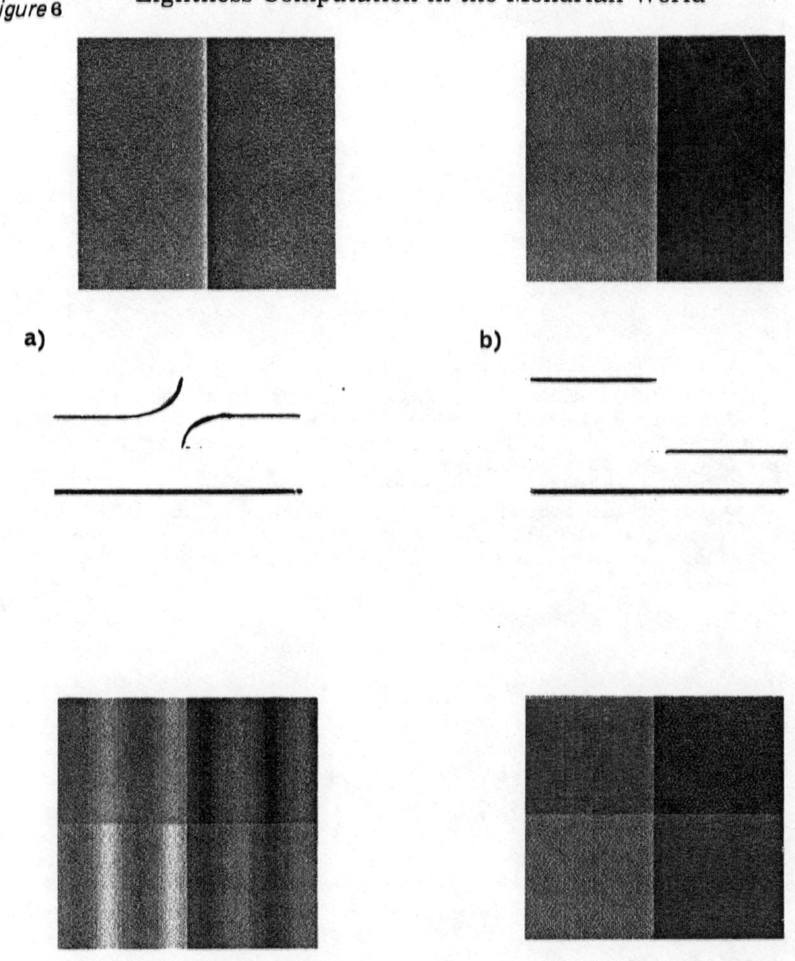

6. Two examples of the performance of the lightness computation described in this paper, and implemented on a serial computer. The Craik-Cornsweet-O'Brien illusion (a) would be generated by the intensity whose <u>logarithm</u> is shown (with intensity profile plotted below) and the output of the computation (b) consists of two regions of uniform lightness, as expected. The log-intensity profile (c) is generated by a piecewise constant reflectance with sinusoidally varying illumination; the resulting lightness (d) is piecewise constant - the sinusoidal variation has been suppressed.

$$l_{i,j}^{(n)} = \sum_{k,m\in N(i,j)} l_{k,m}^{(n-1)}/M(i,j) + (1/2)r_{i,j}$$

where $M(i,j)$ is the number of neighbours of i,j - members of $N(i,j)$. It can also be shown that, when Jacobi relaxation is used, the scaling of the lightness signal is kept under control in this sense: in successive iterations, the spatial average of the lightness stays constant. However, as the diagonal dominance of the iterated operation is not strict, a minor modification is required to ensure convergence (Blake 1983 p. 139).

4. CONCLUSION

This paper has presented a significant refinement of the work of Horn (1974) on the computation of lightness. The method presented here is capable of recovering the reflectance of any Mondrian image. It has been shown that correct treatment of the image boundaries is critical - inappropriate boundary conditions can cause large errors in reflectance recovery. The computational mechanism required bears some similarity to that of Horn's method and so the argument of Marr (1974) for biological feasibility may continue to apply, although Land (1983) has recently argued that lightness computation is not performed in the retina (as Marr suggested). Moreover, because the method performs a global minimisation, it is inherently robust in the presence of local error or noise.

It must be emphasised that the assumption that the image is a Mondrian radically simplifies the lightness computation problem. This is because the detection of reflectance edges becomes trivial and reflectance changes are guaranteed to be separable from illumination changes. This is not, in general, true of natural images.

Acknowledgement

The author acknowledges the invaluable assistance of Dr P. Mowforth, Prof. J. Frisby, S. Pollard, Dr A. Gilbert and R. Tobin. He is also grateful to the University of Edinburgh for the provision of facilities.

References

1. Blake,A. (1983). *Parallel computation in low-level vision*. PhD Thesis, University of Edinburgh.

2. Clemmow,P.C. (1973). *An introduction to electromagnetic theory*. Cambridge University Press.

3. Horn,B.K.P. (1974). Determining lightness from an image. *Computer Graphics and image processing*, 3, 277-299.

4. Land,E.H. and McCann,J.J. (1971). Lightness and retinex theory. *J. Opt. Soc. Am*, 61, 1-11.

5. Land,E.H. (1983). Recent advances in retinex theory and some implications for cortical computations: Color vision and the natural image. *Proc. Natl. Acad. Sci. USA*, 80, 5163–5169.

6. Marr,D. (1974). The computation of lightness by the primate retina. *Vision Res.*, 14, 1377–1388.

7. Marr,D. (1982). *Vision*. Freeman, San Francisco.

8. Strang,G. and Fix,G.J. (1973). *An analysis of the finite element method*. Prentice-Hall, Englewood Cliffs, USA.

Opponent-Colours Theory

L.M. Hurvich

Department of Psychology, University of Pennsylvania, Philadelphia, Pennsylvania 19104, USA

The latter-day Cantabridgians working in vision continue to remind us of the brilliance of Newton, Young and Maxwell in whose shadows they continue to ply their trade. (Mollon and Sharpe, 1983). The contributions of these geniuses are universally admired and understandably so. The list could be extended by adding still another extraordinary Cantabridgian, Lord Rayleigh. Or still another, but this one from the other Cambridge, namely Cambridge, Massachusetts and one whose name as an eponym is virtually a household word to those in this lecture room.

This brilliant visual scientist qualifies as a physicist, a biophysicist, a biochemist, a psychologist, an electrical and mechanical engineer and to top it off an inventive genius. Granted dozens of patents with 234 claims in 1932 for his work in color cinematography, it was he who invented the multicolor film known as Technicolor and perfected both the taking and projecting cameras. Until his untimely death in 1932 at the age of 43, Leonard Thompson Troland, was also a member of the Department of Psychology at Harvard which he had joined in 1916. He had written some 40 papers, several classics among them, and books entitled The Nature of Matter and Electricity (with D. F. Comstock, 1917), The Mystery of the Mind (1926), The Fundamentals of Human Motivation (1928) and his great three volumes The Principles of Psychophysiology (1929, 1930, 1932). The unit of retinal illuminance has been most fittingly named in his honor, the "troland" and is in everyday use in laboratories everywhere. The National Academy of Sciences now administers the Troland Award "To recognize and encourage the research of young investigators in psychophysics," with funds derived from his magnanimous bequest. It seems appropriate to quote Troland on color vision theory.

In volume II of the "Principles," the volume subtitled "Sensation," Troland tells us that a theory of color vision concerns "the retinal and optic nerve correlatives of hue and saturation" and that the available psychophysical facts are variously interpreted "in attempts to guess at the hidden neurological mechanism and its manner of association with visual experience." In characterizing this enterprise he says, "Probably more ingenuity has been expended, with minimum fruitfulness, (in the domain known as the "theory of color vision") than in any other field of scientific speculation."

What he himself then elected to do in his discussion of the problem was to try to render the most prominent theories like those of Young-Helmholtz, Hering and Ladd-Franklin more realistic by putting them in his own words "in modern dress." "This seems a somewhat safer thing to attempt," he wrote, "than the formulation of another "new theory of vision." His caveat still applies and his discussion remains conceptually sound even at this distance. One is surprised to find him suggesting, for example, that the chromatic and achromatic "unit" areas are of different sizes and even venturing a guess about their relative dimensions. In dealing with the Hering hypothesis he says it is difficult to accept it in the light of modern views regarding neural processes but he doesn't scornfully reject assimilation and dissimilation as "meaning nothing" as Hecht did (1934). We find Troland saying instead that there are chemical processes that are enhanced by certain wavelengths and inhibited by others, but that they are not very common.

But, of course, the "modern dress" of the 1930's is long outdated. We all know what a great explosion of knowledge has occurred in the areas of anatomy and histology of the visual mechanism, photochemistry, the electrophysiology of retina, LGN, and cortex, and the chemistry of neural transmission. We will hear about some of the most recent work in some of these areas at this meeting and in the context of ongoing psychophysical studies will try, as I understand it, to put it all together before we disband.

Mention of Troland's attempt to render Young-Helmholtz and Hering in modern dress prompts me to call your attention to an historical oddity. Thomas Young and Helmholtz both rejected the original well known Thomas Young notion of 1802, Young implicitly and Helmholtz explicitly but temporarily. For the Young story I refer you to a paper entitled "Historical Survey of Color Vision Theory" presented by Dorothea Jameson at a Symposium held in March of 1981 at the University of Pennsylvania. The published version of this Symposium is expected to be out in the near future. It is widely known that Young changed the three basic color sensations from red, yellow and blue, to red, green, and violet but it is little appreciated that he changed the theory along with the names. Helmholtz' explicit rejection was made in a 1852 paper. Although Helmholtz later came about, Young himself apparently never did. For the Helmholtz story I refer you to our 1949 paper where the details of the rejection are spelled out (Hurvich and Jameson, 1949).

To top this off it seems curious that there is so little awareness of Helmholtz' appreciation of Hering's theory as a three-variable one, and that Helmholtz linked the two theories mathematically. "E. Hering's much discussed theory," wrote Helmholtz in 1896, "is a modification of Young's theory, which seeks through the choice of other Grundempfindungen, to be in better agreement with what it believes must be regarded as the immediate facts of introspection." We pointed out in our Teddington Symposium paper (Hurvich and Jameson, 1958) that Helmholtz with three linear equations, showed how the three paired variables of the Hering theory related mathematically to his own three-variable schema. As he wrote them they were: $u=\sqrt{1/3}\,(x+y+z)$; $v=\sqrt{1/2}\,(x-z)$; and $w=!\sqrt{6}.(x-2y+z)$, where u, v, and w, represent the white, yellow-blue and red-green dimensions. (Positive v is yellow, negative v is blue; positive w is purple red, negative w is green.)

I have been asked to speak about opponent-color theory. I might have chosen to talk about a variety of opponent models that have been proposed and compared and evaluated them against our own work (Jameson and Hurvich, 1955, 1968, 1972; Jameson, 1972). Among those who have proposed alternative models are Boynton (1960), Peter Walraven (1962), Judd and Yanemura (1970), Guth (1970), Ingling (1977), Massof (with his colleague Bird) (1978), and Guth, Massof and Benzschawel (1980). But rather than evaluate the strengths and weaknesses of the various models, for which there would be no time, in any event, what I shall do instead is to follow up the paper I presented at the 1977 AIC meeting with the title "Two Decades of Opponent Processes" (1978). Now I have a mere six years or so to look at and review. The shortened time scale, I am sure, will be seen later to have given me a distorted perspective.

At the time of the AIC '77 review I was able to report only a single replication of our basic cancellation experiment that had been carried out some 20 years earlier in 1955 (Jameson and Hurvich). The replication appeared in Dr. Romeskie's doctoral dissertation in 1976 and was carried out in connection with her study of anaomalous color vision. A published report did not appear until 1978. A second thesis report by Donnell followed Romeskie's but still remains unpublished (1977). He, like Romeskie, reported sets of cancellation functions very similar to our original ones even though he changed the experimental procedure somewhat. Given the known variability among individuals in their color mixture data it comes as no surprise to find that there are individual differences in the cancellation functions, some quite marked, but there is no question about the robustness of the phenomenon.

We now have additional sets of cancellation data from Werner and Wooten for 3 observers (1979a) as well as those of Ikeda and Ayama of last year (1983). Werner and Wooten (1979b) have also published average chromatic response functions from three of the above cited studies for 7 observers. Using an iodopsin nomograph with lambda max at 435, 530, and 560 nm they have evaluated the weighting functions necessary to apply to the alpha, beta and gamma photopigments to predict the opponent response functions. A least-squares technique was used for these determinations and based on the average chromatic functions Werner and Wooten have also derived a theoretical hue-naming function for the "average observer." Whether average data or individual data are used for computational purposes will obviously depend on the type of question that is being asked. Linearity was assumed in establishing the average functions and I shall return shortly to this still unresolved issue.

Replications aside, the validity of the cancellation method as a measure of hue strength has itself recently been questioned. According to Ingling and his associates (1978) the cancellation method overestimates the red chromatic response magnitudes in the short wavelength region. This happens, they claim, because there is a silent surround receptive field which is an inhibitory surround and it becomes effective only when the excitatory center is active. Thus in the cancellation situation they argue that the activation of the silent surround reduces the effectiveness of the green cancelling light hence the overestimation of the cancelled redness. The appropriate procedure to measure these chromatic responses, according to them, is to use a matching technique. With a mixture of 480 and 680 nm as primaries, matches were made to the spectral lights of the shortwave region. Although it seems unlikely on simple inspection that one could meet a "same amount of redness" criterion in such a matching situation, Ayama and Ikeda (1982) put this to the test. They asked five observers to use an equal redness criterion in one experiment and a hue matching criterion in a second experiment for otherwise identical conditions. They report that it was impossible to apply the former end point (equal redness) but that all observers found the hue matching criterion easy to use. Their further analysis of the hue matching situation in colorimetric terms showed quantitatively that the form of the measured red-green response function is dependent on the choice of the longwave primary. The use of 680 nm as the red primary provides a desaturated mixture with a small amount of redness but as the red primary is moved along the purple line (in the CIE chart) towards violet the shortwave red limb of the red-green function is greatly magnified. The Ingling et al. procedure fails to disconfirm the validity of the cancellation method.

Still another disconfirmation is reported by Ebenhöh and Hemminger (1981). They scaled hue and saturation using direct estimation techniques and report discrepancies between their transformed curves and the Hurvich and Jameson cancellation functions. It is hard to give this report serious weight since it stands in direct conflict with Werner and Wooten's work. In the latter experiments both the cancellation data and the hue scaling data were obtained on the same observers and the two sets of data for each individual were directly compared. "The results showed that the hue scaling was accurately predicted from the cancellation functions..." (Werner and Wooten, 1979a).

A somewhat different but related issue is raised by Krauskopf et al. (1982). They determined in a chromaticity diagram what they call the cardinal directions of color space using a selective adaptation technique with temporally varying lights to measure threshold color and luminance changes. They report that their experiments generally support the opponent color theory but that the results conflict "with Hurvich and Jameson's analysis in an important respect, namely by identifying one of the cardinal chromatic axes with a tritanopic confusion line and not with a unique yellow-blue (red-green equilibrium) line."
What they have really shown is that the major chromaticity axes can be defined in at least two different ways: as unique hue axes or as maximum desensitization axes. In their experiments the red-green unique hue axis corresponds to one maximum desensitization axis but the yellow-blue unique hue axis deviates from a second maximum desensitization axis. Their results would seem to be consistent with desensitization occuring at the cone rather than at the opponent level. Since Krauskopf et al. have provided us with no a priori definition of "cardinal axes" and since the different definitions do not agree it would probably be well not to use the term "cardinal axes."

A basic premise of the opponents notion is that the members of an opponent pair of colors are mutually exclusive. The cancellation experiments are based on this premise. Red percepts exclude green ones and vice versa. Yellows exclude blues and the same for blacks and whites. Red and green on the one hand and yellow and blue on the other are never seen simultaneously in the same place at the same time in a color (Hering, 1878).

This is a generally accepted fact among visual scientists and carefully executed hue-naming experiments have always supported this generalization (Jameson and Hurvich, 1959; Boynton and Gordon, 1965; Werner and Wooten, 1979a). Perceptual analysis is nonetheless not externally verifiable and one sometimes encounters students--and even professors--who claim to see both yellow and blue simultaneously in green hues. These apparent falsifications of a basic premise of the theory can usually be traced to an individual's experience with pigment color mixtures.

Crane and Piantanida (1983) now report an experiment in which they stabilized the retinal image of the boundary between a pair of red and green stripes but not their outer edges and when they did so their observers saw one or more of three distinctly different appearances. The field appeared to be composed entirely of a regular array of just resolvable red and green dots, or as a series of islands of one color on a background of the other or the entire field appeared to be a single unitary color composed of both red and green--a field described as simultaneously red and green.

Some observers reported it as colored but unnameable, and still others as undescribable. Similar results were obtained with yellow and blue by some observers. Although some readers reacting, I presume, to the title of the paper which is "On seeing Reddish-Green and Yellowish-Blue" seem to believe that Crane and Piantanida are arguing that the exclusivity concept has been invalidated, they are in fact not doing so for normal viewing conditions. What they do suggest is that the percepts reported are related to the filling-in phenomena commonly obtainable under stabilized image conditions and that we may be dealing with cortico-cortical processing rather than with retino-cortical color processing in the primary visual pathway.

In this context, it should be noted that there has been a long history of disagreement in the literature about the white-black system. Although the opponency of white and black are obvious in both successive and simultaneous contrast, black and white sensations have been widely held to mix to produce grays in a way that the red-green and yellow-blue systems do not mix to produce an intermediate mixture result. Perceptually one seems to perceive degrees of similarity between grays and their presumed black and white constituents. We have never regarded this as a problem. Mid-gray has not been treated by us as a mixture of black and white. Our conception of mid-gray is that it is the intrinsic basal sensation associated with the equilibrium condition of the entire visual system. Non-balanced excitations in the paired visual response systems can then be related to various degrees of departure from the basal mid-gray state. In the white-black system such unbalanced excitations are related to departures from mid-gray toward either the whiteness or blackness direction in the same way that they occur for the yellow-blue and red-green systems. And now, the electrophysiology data can be appealed to in buttressing the opponency notion as it applies to black and white.

In any event we seem to have come a long way since I discussed the "Indispensibility of a Bimodal Black-White Color Vision Process" in a 1966 paper. Even W. D. Wright recently appears to have independently discovered the importance of the blackness sensation. This is reported in a note in Leonardo (1981).

In our own modeling of the opponent systems we have treated the white function as identical with the measured luminosity function and have usually shown it graphically in published figures along with the paired chromatic systems. No measures of the blackness function have been available but in one of our papers we wrote: "The induced, rather than the directly stimulated, black component of the achromatic white-black response pair is not shown in the figure. It would have the same distribution as the white but would be of opposite sign, since the strength of the black contrast response is directly related to the magnitude of either the surrounding or the preceding whiteness or brightness." (Hurvich, 1960).

Thanks to Werner, Cicerone and their colleagues (Werner et al., 1984; Cicerone et al., 1984) we now have measures of the blackness function. In one of their experiments a small foveal white field (45') was centered in a monochromatic annulus (60'-120') and separated from it by a dark surrounding gap. The radiance of the annulus was adjusted until the white field just turned black and was indistinguishable from the dark gap. Assuming that the radiance required to induce blackness is inversely proportional to the sensitivity of the blackness function this curve agrees with the spectral sensitivity function determined by heterochromatic flicker photometry for the annulus. Data are available for four observers and for retinal illuminances ranging from 50-1200 trolands. In a second experiment with essentially the same stimulus arrangement but using central areas of different wavelengths, i.e., 480, 500, 580, or 660 nm the measured efficiency of induced black still matched the inverse of the CIE V_λ function. Control experiments measuring increment thresholds under the same conditions matched neither the V_λ function nor the blackness function. The authors claim also that the chromatic pathways do not contribute to the perception of blackness. However, the question of the contribution of the chromatic processes to brightness has been a long debated one and if they do then it is hard to see how they would fail to contribute also to blackness in an induction situation. I shall not discuss the relation of the chromatic systems to brightness here (Yaguchi and Ikeda, 1983).

By inducing some blackness with adjacent bright surrounds into spectral yellows and reds it is easy to demonstrate that they are changed into browns (Hurvich, 1981). If yellow and red opaque pigment colors are desaturated with blackness we achieve the same result. Despite these simple psychophysical demonstrations and their obvious physiological implications there is a persistent tendency to treat brown as a distinctive color with the property of uniqueness akin to the perceptually unique hues red, yellow, green and blue.

Fuld, Werner, and Wooten (1983) have pursued this issue recently with a variation of the paradigm used by Sternheim and Boynton (1966) to demonstrate that orange is not an elemental hue in the same sense that red and yellow are. Fuld et al. applied the percentage hue scaling technique to evaluate the appearance of a single wavelength, say 580 nm as the surround illumination was increased. Different sub-sets of color categories from the set that includes yellow, red, black, white and brown were used in different experiments and the conclusion reached that the 580 nm test field required the color name brown in addition to the names yellow and black when intermediate surround intensities were used. Their conclusion is somewhat guarded. The title of their paper is "The Possible Elemental Nature of Brown." Follow-up work in Dr. Wooten's laboratory seems to have increased the "guardedness." It is clear that the instructions given the observer, the observer's response biases, the specific method, and particularly the order in which the color categories are made available to the observer in a series of experiments can all be shown to influence the results (Wooten, 1983).

Bartelson, a few years earlier (1976), had examined experimentally some of the issues relating to brown and sought answers to the questions: what is brown, and is brown unique? Using direct scaling techniques with a variety of Munsell color chips he showed that brown was a three-dimensional color as distinct from a one dimensional hue perception. The conditions necessary to elicit a brown response may be specified according to relationships among three dimensions of color. If we go to Webster's 3rd New International Dictionary, as Bartelson points out, we have a sound definition. Brown is "any of a group of colors between red and yellow in hue, of medium to low brightness, and of moderate to low saturation." It does not have the same property of uniqueness associated with the unitary hues red, yellow, green, or blue.

In connection with the induction issue I would like to recall our generalized formulation for color (Jameson and Hurvich, 1959; Hurvich and Jameson, 1960). Induced effects are treated as increments or decrements in the response activity in the focal area and our formal definition of perceived color (C) is the following:

$$C = f[\Sigma(e_\lambda \chi_\lambda), \Sigma(e_\lambda \psi_\lambda), \Sigma(e_\lambda \Omega_\lambda)] + I$$

where e_λ is the stimulus energy, at any visible wavelength in the focal area, $\chi_\lambda, \psi_\lambda, \Omega_\lambda$ are the relative sensitivies of three

generalized response variables of the visual color system, and I represents the total induced response activity. This relation may be written more specifically in terms of the three variables of the opponent colors theory as:

$$C = f[(y\text{-}b)_f, + (y\text{-}b)_i, (r\text{-}g)_f + (r\text{-}g)_i, (w\text{-}bk)_f + (w\text{-}bk)_i]$$

where the focal terms are responses of the three paired systems produced by the direct stimulation of the focal area, and the induced terms are the responses induced in the corresponding focal area by stimulation of neighboring regions of the retina (Hurvich and Jameson, 1960; Jameson and Hurvich, 1964).

At the time of the Color '77 review the red-green opponent system without exception was recorded as behaving linearly based on the results of a number of different investigators (Hurvich, 1978). Unique blue and unique yellow had been shown to be invariant with luminance changes and linearly additive, i.e., lights that are

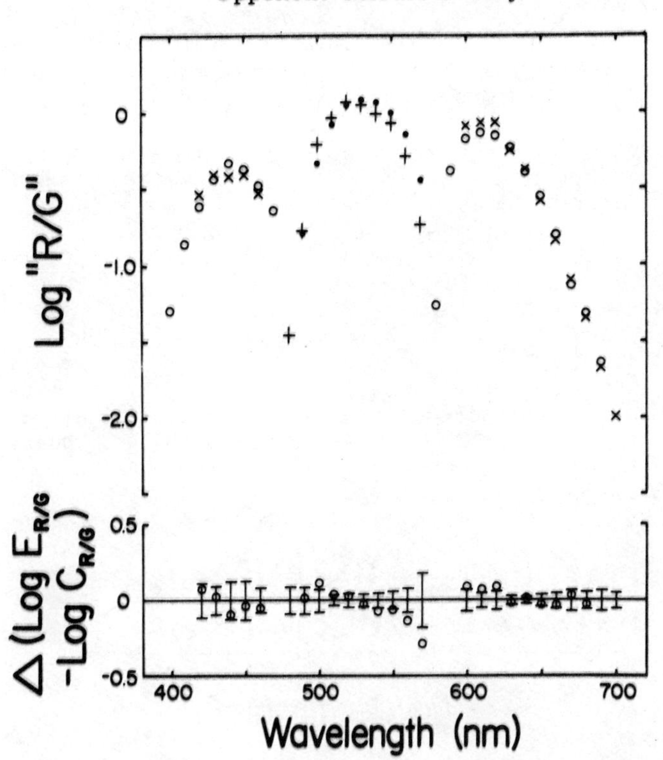

FIGURE 1. Red-Green Cancellation Data. Observer H. See text.

uniquely blue when combined with lights that are uniquely yellow resulted in mixture lights that were uniquely blue, uniquely yellow or achromatic whatever the luminance ratios of the components. Additional data supporting red-green linearity come from Werner and Wooten (1979a) who have fit the red-green chromatic response functions of their three observers with a linear combination of three photopigments and also from Ayama and Ikeda (1980).

The yellow-blue opponent system, on the other hand seems to be a more variable one and is more frequently found to be non-linear than linear (Hurvich, 1977; Werner and Wooten, 1979a). Some individuals, however, do show yellow-blue linearity (Romeskie, 1978; Werner and Wooten, 1979a). Our own data can be fit reasonably well with linear combinations of iodopsin type photopigments. The results of the two observers are shown in Figures 1, 2, 3 and 4. The different symbols represent the experimental data, on the one hand, and the computed linear fits, on the other. The lower part of the graph shows the log difference between the calculated value and each experimental measure along with the associated error bars.

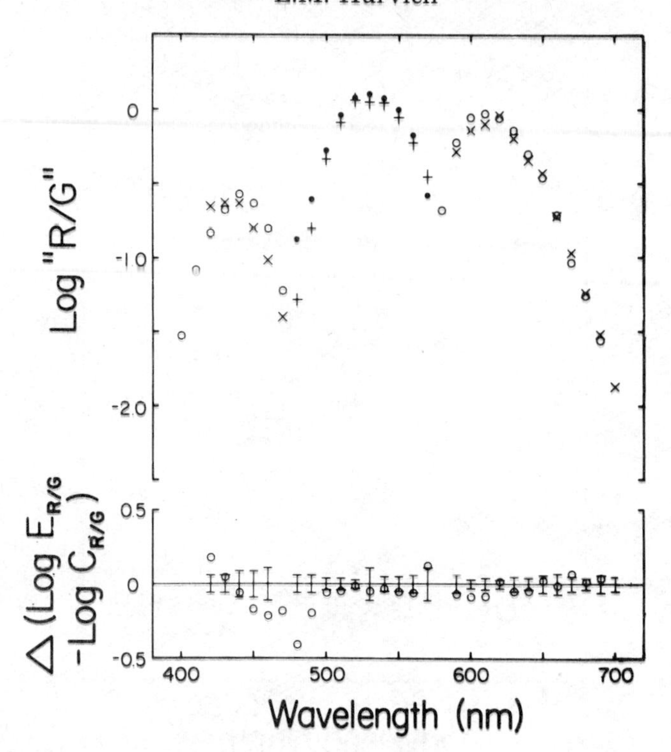

FIGURE 2. Red-Green Cancellation Data. Observer J. See text.

Ikeda and Ayama (1980, 1983), make use of a summation index to evaluate the linearity of the various chromatic systems. They first find the cancellation radiances of two wavelengths from the same spectral region necessary to achieve equilibrium with a single reference stimulus of opposite sign. Then with mixtures of the two cancellation stimuli in different proportions another equilibrium check is executed.

Additivity is found for green (also for red if the pairs of stimuli are selected from the same spectral region) and for blue. The results for yellow indicate non-additivity supporting the view that the yellow-blue opponent system is non-linearly related to the cone inputs, although Ikeda and Ayama say that no satisfactory non-linear model has been proposed.

The very recent report of Burns et al. in Vision Research (1984) stands by itself. They compared unique and other constant hue loci under similar conditions for two observers. Expressed in Judd chromaticity coordinates unique hue and constant hue data agreed. The unique blue loci were curved and unique red and green loci were not colinear. These data imply that unique hues are not a

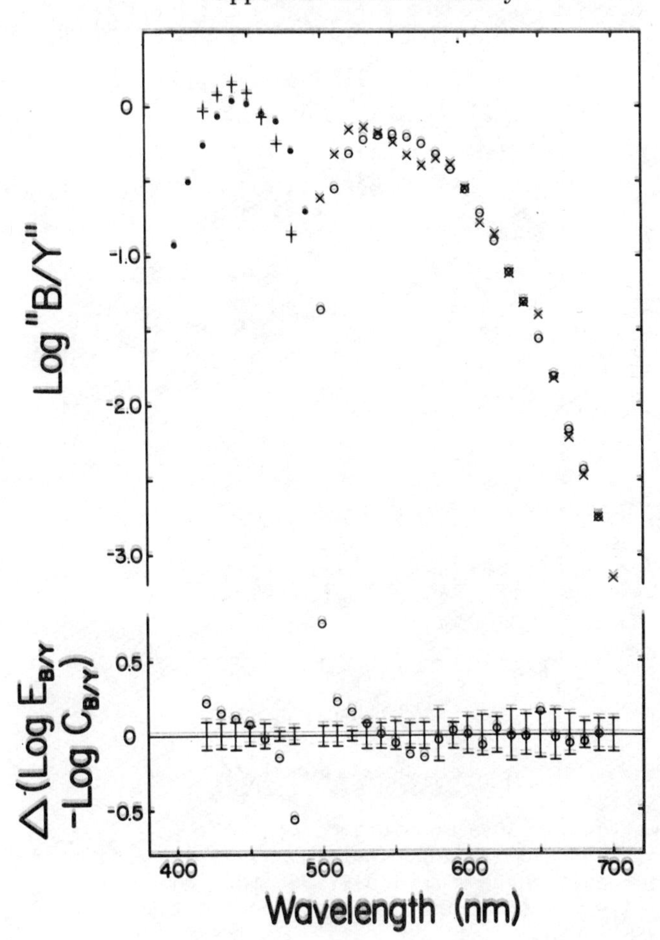

FIGURE 3. Yellow-Blue Cancellation Data. Observer H. See text.

linear transformation of the color matching functions and hence all linear models are only an approximation even at a single luminance level. They conclude that both the red-green and yellow-blue chromatic systems are non-linear.

This report will bear close examination. The use of "white light" as one of the mixture stimuli means that the role of macular pigmentation will have to be looked into. Furthermore they report that desaturating a unique blue with the observers own unique yellow made it look reddish. This is quite puzzling on the face of it and hard to reconcile with the Larimer et al. data (1974). The fact that the spectral unique blue also looked reddish when desaturated with 571 nm which the authors tell us was a slightly

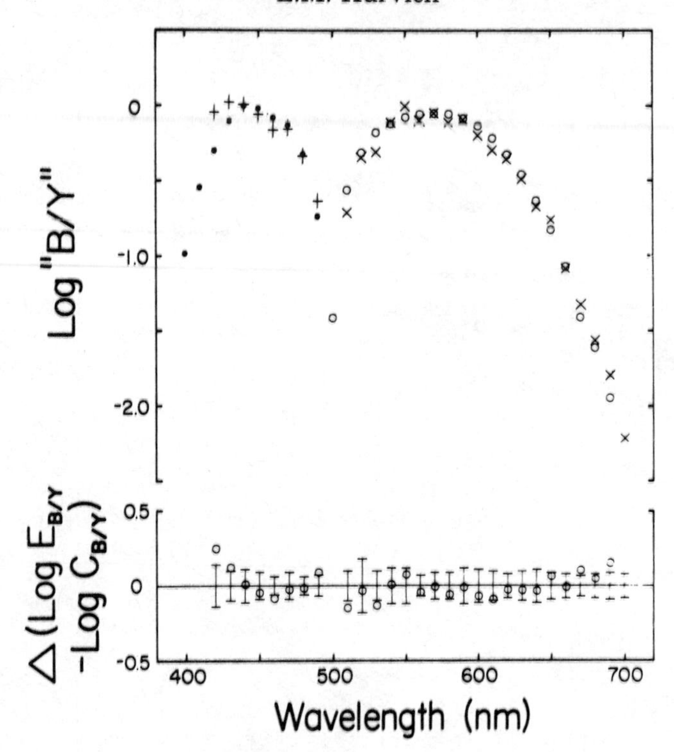

FIGURE 4. Yellow-Blue Cancellation Data. Observer J. See text.

greenish-yellow is cause for further caution. There is an implication in these findings that the observers were obtaining their unique hues for non-neutrally adapted conditions. A biased adaptive state can eventuate in unanticipated hue remainders. This remains to be seen.

Color deficient individuals who lack red-green chromatic response systems (Hurvich, 1972) i.e., protanopes and deuteranopes have an achromatic spectral locus, a neutral point. In every test of the behavior of their yellow-blue system the outcome has been the same: it behaves linearly. Ikeda and Ayama (1983) have checked the additivity of a protanope's yellow response function with the summation index technique and for a wavelength combination of 530 and 630 nm linear summation was found. His yellow-blue cancellation functions were satisfactorily fitted with a linear combination of the standard color matching functions. Knoblauch and Wooten (1982) have recently checked the achromatic points of two protanopes and two deuteranopes over a retinal illuminance range of 4 log units and found them to be invariant. Starr (1978) reported a similar result over a somewhat smaller range and Romeskie and Yager (1978) obtained yellow-blue hue cancellation data from one protanope and one dueteranope which were well fit by a linear combination of iodopsin extinction spectra.

Apparently the yellow/blue non-linearity, when it occurs, is related to the condition of more than one cone input into the yellow lobe.

At the AIC meetings in Luzern in 1965, Jameson presented a paper on threshold and supra-threshold brightness relations for different stimulus durations. She reported, among other things, that the fovea was less sensitive than the periphery for the experiments under consideration. Furthermore, foveal brightness rose more steeply as a function of stimulus intensity than it did in the periphery. These facts were discussed in the context of the recruitment phenomenon which occurs in diverse sensory domains. It is best known in audition as "auditory recruitment." It is a phenomenon associated with certain types of hearing loss in which the loudness of tones appear to increase more rapidly than normal. It seems to be true that if threshold sensitivity differences are related to differences in density or availability of neural response elements, an increase in threshold, or sensitivity loss, seems to be associated with a more rapid increase, a steeper slope, in the psychophysical function that relates the sensory response to stimulus magnitude. In the chromatic system specific threshold energies for red and green hue responses to spectral stimuli are lower than the thresholds for yellowness and blueness, yet the yellow and blue hues increase in strength with increase in luminance at a more rapid rate than do the red and green, thus giving rise to the Bezold-Brücke hue shifts (Hurvich and Jameson, 1958).

In view of the well known zonal properties of the eye, namely that the yellow-blue zone extends beyond the paired red-green zone we postulated a reversal of the Bezold-Brücke hue shift in the periphery. With several of our students we spent a good deal of effort on the problem but could never satisfy ourselves about the reliability of our results because of their highly variable nature. Peripheral color matching is difficult and made more so by fading (Troxler effect). It is even more difficult when hue matching is attempted across different luminance levels.

We are delighted to report that Stabell and Stabell (1982) have cracked the experimental problem. They made their measurements at a series of peripheral locations during the cone-plateau period of the long-term dark adaptation curve, a technique which they have used in a number of earlier studies. They report that the Bezold-Brücke hue shift was found to reverse direction in going from the central fovea into the far peripheral retina; in the periphery, "an increase in the luminance level in the middle- and long-wave regions of the spectrum produces, respectively, a trend toward green and red instead of toward yellow."

I believe that in the past few years the opponent process theory has raised more questions and stimulated more excitement and more activity in the area of adaptation than in any other area. The coming together of the Stiles' π-mechanism work (Stiles, 1959,

1978) and the "two process" interpretation of the phenomena of adaptation (Jameson and Hurvich, 1961, 1972) seem to have released a remarkable burst of energy in the field related to the testing of a variety of specific hypotheses. We have had increment threshold experiments, cancellation experiments, luminance vs. chromatic threshold discrimination analyses, etc. Among those who are associated with experimental work of this sort are Foster, Guth, Hood, Ingling, King-Smith, Larimer, Loomis, Mollon, Pugh, Reeves, Shevell, Stromeyer and Wandell (Mollon and Sharpe, 1983). Dr. Pugh will discuss experiments and issues that fall in these categories. I will mention only one matter of interpretation. Hood and Finkelstein (1983) have interpreted the results of some of these experimental manipulations as implying that the so-called tuning of particular classes of cells is variable rather than a fixed characteristic.

We have long seen the need in the analysis of adaptation mechanisms to involve both the receptoral and post-receptoral levels (Jameson and Hurvich, 1961, 1964, 1972). The recent work in our laboratory has dealt with the separate recovery processes at the two levels (Jameson et al., 1979). We did this by tracking the spectral locus of unique yellow during recovery from chromatic adaptation to either long wave or middle wave spectral lights. Two temporal exposure conditions were used: either 5 minutes of steady viewing or 5 minutes of total light exposure distributed over a 10 minute period in a repeated sequence of a 10 sec. exposure to the adapting light followed by a 10 second dark interval. Compared to the recovery process after a steady adaptation, the alternating dark and spectral light exposures produce initially smaller unique yellow shifts but the recovery process is slower. Whereas a single exponential describes recovery in the first case, recovery in the second instance is described by a sum of two different exponentials. It is clear that the rapid desensitization and recovery at the cone level serves as a protective mechanism for the neural elements beyond. it is also our hunch that the long term post-receptoral biasing effect produced by temporal cycling is a crucial aspect of habituation phenomena such as the McCollough Effect (1965).

Although there is a vast literature related to flicker phenomena this did not keep us from entering the lists with a paper on temporal sensitivities published with our doctoral student D. Varner in the May issue of the Journal of Optical Society of America (Varner et al., 1984).

I am not going to try to summarize the experimental results. I want only to say that we have reexamined some aspects of sensitivity to temporal variations in luminance and chromaticity in terms of opponent-process theory. We have explored a number of different questions such as those related to individual differences in the forms and levels of temporal sensitivity functions for luminance and chromaticity variations, addressed the question of the linearity of the red-green system when it is temporally modulated and looked at the effects of adapting backgrounds that are either in red-green equilibrium or not.

Jameson will discuss some other recent work on chromatic factors in threshold and suprathreshold spatial vision in her paper.

I do not have time to talk about recent work in color deficiencies as I did at AIC '77 but I would like to call your attention to an aspect of opponent-colors theory that predates and continues into the time period I am reviewing. Long before computational models of the visual system became of interest to physiologists, one of our goals in quantifying opponent-colors theory was to specify color appearance in a coordinate system that would represent the three variables of perceived color (hue, brightness, and saturation) associated with the spectral qualities, quantities and spatial distributions of retinal light images, by way of specified relations to the retinal receptor signals and their processing through the three-variable opponent system.

The derived HBS system enabled us to specify the appearance of the spectrum quantitatively for different conditions of receptor adaptation. The derivation was based on the measured chromatic and achromatic functions and their computed changes with changes in the balance of receptor sensitivities. These specifications are summarized in the color diagram shown in Figure 5. At a constant brightness level, the hue coefficients are percentages of the chromatic responses and are plotted as angles in polar coordinates in the hue circle. The saturation coefficient is expressed along the radii where zero saturation lies at the center of the circle and maximal saturation lies on the circumference.

Essentially the same procedure is used to specify the color of any object of known spectral reflectance in the color diagram. We need only calculate the cross products of the reflectance of the surface, the relative light energy distribution of the illuminant, and the chromatic and achromatic responses for a specified condition of adaptation. The cross-products are then simply summed over wavelength and hue and saturation coefficients are expressed in percentage terms just as in the case of spectral stimuli. Calculated values of these coefficients for broad-band reflectance (or transmittance) samples all lie within the spectral locus shown in Figure 5 (Hurvich and Jameson, 1956; Jameson and Hurvich, 1967; Hurvich, 1981).

In relating the physical characteristics of chromatic stimuli to the perceptual attributes of the visual response we made it clear in our early papers that we were limiting ourselves to a single luminance, hence a single brightness level. Computations were presented for a number of different colored light adaptations using a v.Kries receptor desensitization analysis, and various

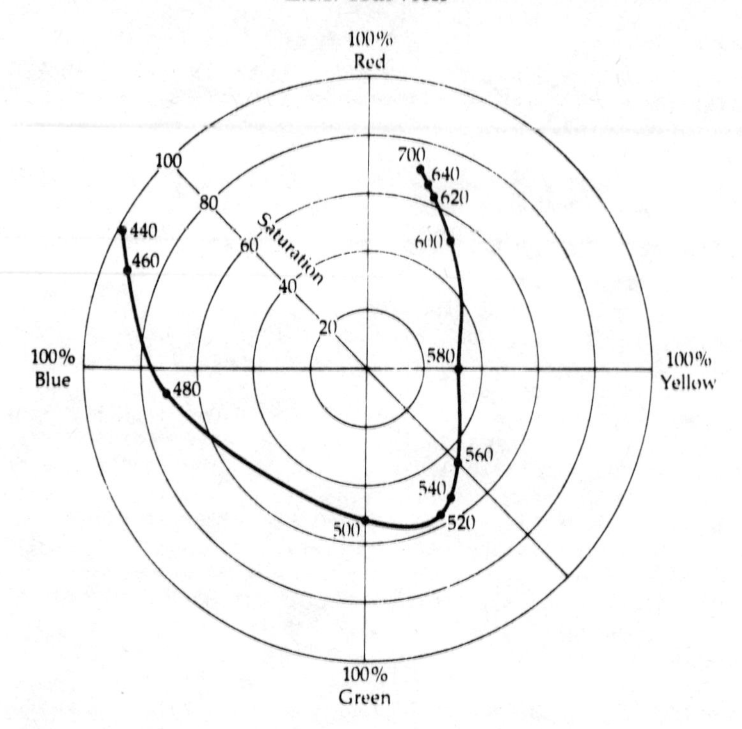

FIGURE 5. Hue and Saturation Coefficients of Spectral Stimuli Represented in a Polar Coordinate Diagram. Hue is Plotted Circumferentially and Saturation Along the Radius. This is a Perceptual Space. The Brightness Dimension is orthogonal to Plane of the Page.

object color specifications were derived for various conditions of illumination and adaptation.

Our continued development of this approach included the induction factor that I referred to in the generalized definition of color I presented above. The dependence of these color specifications on the spatial dimensions of the stimuli and their relation to receptive field properties entered as a still further development and will be discussed by Jameson in her paper.

I'll conclude this review with a preview. Dr. David Krantz whose elegant formal treatment of color theory is known to you all, has sent us a draft manuscript dealing with constraints on theories of neural mechanisms of color. The paper is called "The Topological Laws of Color Matching" and will probably appear in the Journal of Mathematical Psychology. I mention it in closing because of the impact made by his earlier formal treatment of color and color theory.

REFERENCES

Ayama, M. and Ikeda, M. (1982). Chromatic Valence Curves: Alternative Interpretation Derived by the Direct Matching Method. Science, 215, 1538-1539.

Bartelson, J.C. (1976). Brown. Color Res. Applic. 1, 181-191.

Bird, J.F. and Massof, R.W. (1978). A General Zone Theory of Color and Brightness Vision. II. The Space-Time Field. J. Opt. Soc. Am., 68, 1471-1481.

Boynton, R.M. (1960). Theory of Color Vision. J. Opt. Soc. Am., 50, 929-944.

Boynton, R.M. and Gordon, J. (1965). Bezold-Brücke Hue Shift Measured by Color Naming Technique. J. Opt. Soc. Am., 55, 78-86.

Burns, S.A., Elsner, A.E., Pokorny, J. and Smith, V.C. (1984). The Abney Effect: Chromaticity Coordinates of Unique and Other Constant Hues. Vision Res. 24, 479-489.

Cicerone, C.M., Valbrecht, V.J., Werner, J.S. and Donnelly, S.K. (1984). The Perception of Blackness. Investig. Ophthalmol. & Vis. Science, Suppl. 25, 234.

Comstock, D.F. and Troland, L.T. (1917). The Nature of Matter and Electricity. D. Van Nostrand, New York.

Crane, H.D. and Piantanida, T.P. (1983). On Seeing Reddish Green and Yellowish Blue. Science, 221, 1078-1080.

Donnell, M.L. (1977). Individual Red/Green and Yellow/Blue Opponent Isocancellation Functions: Their Measurement and Prediction. Ph.D. Thesis, University of Michigan, University Microfilms, Ann Arbor.

Ebenhöh, H. and Hemminger, H. (1981). Scaling of Color Sensation by Magnitude Estimation: A Contribution to Opponent-Colors Theory. Biol. Cybernetics. 39, 227-237.

Fuld, K., Werner, J.S., and Wooten, B.R. (1983). The Possible Elemental Nature of Brown. Vision Res., 23, 631-637.

Guth, S.L. (1970). A New Color Model. Proc. Helmholtz Mem. Symp. Color Metrics, Driebergen.

Guth, S.L., Massof, R.W. and Benzschawel, T. (1980). Vector Model for Normal and Dichromatic Color Vision. J. Opt. Soc. Am., 70, 197-212.

Hecht, S. (1934). Vision. II. The Nature of the Photoreceptor Process. In Handbook of General Experimental Psychology. (ed. C. Murchison). Clark University Press, Worcester, Mass.

Helmholtz, H.v. (1852). Ueber die Theorie der zusammengesetzten Farben. Ann. Phys. Chem., 163, 45-66. (English Translation, Phil. Mag. Ser. (4), 4, 519-534, 1852).

Helmholtz, H.v. (1896). Handbuch der physiologischen Optik. 2nd ed. Voss, Hamburg.

Hering, E. (1878). Zur Lehre vom Lichtsinne. Carl Gerold's Sohn, Vienna.

Hood, D.C. and Finkelstein, M.A. (1983). A Case for Revision of Textbook Models of Color Vision: The Detection and Appearance of Small Brief Lights. In Colour Vision: Physiology and Psychophysics. (eds. J.D. Mollon and L.T. Sharpe) Academic Press, London.

Hurvich, L.M. (1960). The Opponent-Pairs Scheme. In Mechanisms of Colour Discrimination. Pergamon Press, New York.

Hurvich, L.M. (1966). The Indispensibility of a Bimodal Black-White Color Vision Process. In Internationale Farbtagung, Luzern. Musterschmidt, Göttingen.

Hurvich, L.M. (1972). Color Vision Deficiencies. In Handbook of Sensory Physiology. VII/4, Visual Psychophysics. (eds. D. Jameson and L.M. Hurvich). Springer, New York.

Hurvich, L.M. (1978). Two Decades of Opponent Processes. In Color '77 (AIC). (eds. F.W. Billmeyer and G. Wyszecki). Adam Hilger, Bristol.

Hurvich, L.M. (1981). Color Vision. Sinauer, Sunderland, Mass.

Hurvich, L.M. and Jameson, D. (1949). Helmholtz and the Three-Color Theory: An Historical Note. Amer. J. Psychol., 62, 111-114.

Hurvich, L.M. and Jameson, D. (1955). Some Quantitative Aspects of an Opponent-Colors Theory. II. Brightness, Saturation, and Hue in Normal and Dichromatic Vision. J. Opt. Soc. Am., 45, 602-616.

Hurvich, L.M. and Jameson, D. (1956). Some Quantitative Aspects of an Opponent-Colors Theory. IV. A Psychological Color Specification System. J. Opt. Soc. Am., 46, 416-421.

Hurvich, L.M. and Jameson, D. (1958). Further Development of a Quantified Opponent-Colours Theory. In Visual Problems of Colour. II. Her Majesty's Stationery Office, London.

Hurvich, L.M. and Jameson, D. (1960). Perceived Color, Induction Effects, and Opponent-Response Mechanisms. J. Gen. Physiol. 43 (6) Suppl. 63-80.

Ikeda, M. and Ayama, M. (1980). Additivity of Opponent Chromatic Valence. Vision Res., 20, 995-999.

Ikeda, M. and Ayama, M. (1983). Nonlinear Nature of the Yellow Chromatic Valence. In Colour Vision: Physiology and Psychophysics. (eds. J.D. Mollon and L.T. Sharpe). Academic Press, London.

Ingling, C.R. Jr. (1977). The Spectral Sensitivity of the Opponent-Color Channels. Vision Res., 17, 1083-1089.

Ingling, C.R. Jr., Russell, P.W., Rea, M.S. and Tsou, B.H.-P. (1978). Red-Green Opponent Spectral Sensitivity; Disparity Between Cancellation and Direct Matching Methods. Science, 201, 1221-1223.

Jameson, D. (1965). Threshold and Supra-Threshold Relations in Vision. In Internationale Tagungsband, Luzern. Musterschmidt, Göttingen.

Jameson, D. (1972). Theoretical Issues of Color Vision. In Handbook of Sensory Physiology. VII/4. Visual Psychophysics. (eds. D. Jameson and L.M. Hurvich). Springer, Berlin.

Jameson, D. and Hurvich, L.M. (1955). Some Quantitative Aspects of an Opponent-Colors Theory. I. Chromatic Responses and Spectral Saturation. J. Opt. Soc. Am., 45, 546-552.

Jameson, D. and Hurvich, L.M. (1956a). Some Quantitative Aspects of an Opponent-Colors Theory. III. Changes in Brightness, Saturation, and Hue with Chromatic Adaptation. J. Opt. Soc. Am., 49, 890-898.

Jameson, D. and Hurvich, L.M. (1956b). Theoretical Analysis of Anomalous Trichromatic Color Vision. J. Opt. Soc. Am., 46, 1075-1089.

Jameson, D. and Hurvich, L.M. (1959). Perceived Color and Its Dependence on Focal, Surrounding, and Preceding Stimulus Variables. J. Opt. Soc. Am., 49, 890-898.

Jameson, D. and Hurvich, L.M. (1961). Opponent Chromatic Induction: Experimental Evaluation and Theoretical Account. J. Opt. Soc. Am., 51, 46-53.

Jameson, D. and Hurvich, L.M. (1964). Theory of Brightness and Color Contrast in Human Vision. Vision Res., 4, 134-154.

Jameson, D. and Hurvich, L.M. (1967). The Science of Color Appearance. Color Engineering, 5, 29-36, 43.

Jameson, D. and Hurvich, L.M. (1968). Opponent-Response Functions Related to Measured Cone Photopigments. J. Opt. Soc. Am., 58, 429-430.

Jameson, D. and Hurvich, L.M. (1972). Color Adaptation: Sensitivity, Contrast, Afterimages. In Handbook of Sensory Physiology. VII/4. Visual Psychophysics. (eds. D. Jameson and L.M. Hurvich). Springer, Berlin.

Jameson, D., Hurvich, L.M., and Varner, F.D. (1979). Receptoral and Postreceptoral Visual Processes in Recovery from Chromatic Adaptation. Proc. Natl. Acad. Sci., 76, 3034-3038.

Judd, D.B. and Yonemura, G.T. (1970). CIE 1960 UCS Diagram and the Müller Theory of Color Vision. In Proc. Int'l. Color Association. Stockholm, Musterschmidt, Göttingen.

Knoblauch, K. and Wooten, B.R. (1982). Intensity Invariance of the Achromatic Point in Sex-Linked Dichromacy. In Docum. Ophthal. Proc. Series, 33. (ed. G. Verriest). W. Junk, The Hague.

Krauskopf, J., Williams. D.R. and Heeley, D.W. (1982). Cardinal Directions of Color Space. Vision Res. 22, 1123-1131.

Larimer, J., Krantz, D.H., and Cicerone, C.M. (1974). Opponent-Process Additivity. I. Red/Green Equilibria. Vision Res., 14, 1127-1140.

Massof, R.W. and Bird, J.F. (1978). A General Zone Theory of Color and Brightess Vision. I. Basic Formulation. J. Opt. Soc. Am., 68, 1465-1471.

McCollough, C. (1965). Color Adaptation of Edge-detectors in the Human Vision System. Science, 149, 1115-1116.

Mollon, J.D. and Sharpe, L.T. (1983). Color Vision: Physiology and Psychophysics. Academic Press, London.

Romeskie, M.I. (1976). Chromatic Opponent-Response Functions of Anomalous Trichromats. Ph.D. Thesis, Brown University. University Microfilms, Ann Arbor.

Romeskie, M.I. (1978). Chromatic Opponent-Response Functions of Anomalous Trichromats. Vision Res., 18, 1521-1532.

Romeskie, M.I. and Yager, D. (1978). Psychophysical Measure and Theoretical Analysis of Dichromatic Opponent-Response Functions. Mod. Prob. Ophthalmol., 19, 212-217.

Stabell, B. and Stabell, U. (1982). Bezold-Brücke Phenomenon of the Far Peripheral Retina. Vision Res., 22, 845-849.

Starr, S.J. (1978). Color Matching Additivity and the Linearity of the Opponent-Chromatic Response Functions in Dichromats. Ph.D. Thesis, University of Pennsylvania. University Microfilms, Ann Arbor.

Sternheim, C. and Boynton, R.M. (1966). Uniqueness of Perceived Hue Investigated with a Continuous Judgmental Technique. J. Exper. Psychol., 72, 770-776.

Stiles, W.S. (1959). Color Vision: The Approach Through Increment Threshold Sensitivity. Proc. Natl. Acad. Sci., 45, 100-114.

Stiles, W.S. (1978). Mechanisms of Colour Vision. Academic Press, London.

Troland, L.T. (1926). The Mystery of Mind. D. van Nostrand, New York.

Troland, L.T. (1928). The Fundamentals of Human Motivation. D. van Nostrand, New York.

Troland, L.T. (1929). The Principles of Psychophysiology. I. The Problems of Psychology and Perception. D. van Nostrand, New York.

Troland, L.T. (1930). The Principles of Psychophysiology. II. Sensation. D. van Nostrand, New York.

Troland, L.T. (1932). The Principles of Psychophysiology. III. Cerebration and Action. D. van Nostrand, New York.

Varner, D., Jameson, D. and Hurvich, L.M. (1984). Temporal Sensitivities Related to Color Theory. J. Opt. Soc. Am., A/1, 474-481.

Walraven, P. (1962). On Mechanisms of Colour Vision. Ph.D. Thesis, Utrecht.

Werner, J.S. and Wooten, B.R. (1979a). Opponent Chromatic Mechanisms: Relation to Photopigments and Hue Naming. J. Opt. Soc. Am. 69, 422-434.

Werner, J.S. and Wooten, B.R. (1979b). Opponent Chromatic Response Functions for an Average Observer. Perc. & Psychophysics., 25, 371-374.

Werner, J.S., Cicerone, C.M., Kliegl, R. and DellaRosa, D. (1984). The Spectral Efficiency of Blackness Induction. J. Opt. Soc. Am. (Submitted for publication).

Wooten, B.R. (1983). Personal Communication. B.A. Honors Thesis, Brown University. J.A. Rosano, III. Biasing Effects of the Continuous Judgmental Color-Naming Technique Applied to Color Brown.

Wright, W.D. (1981). The Nature of Blackness in Art and Visual Perception. Leonardo, 14, 236-237.

Yaguchi, HJ. and Ikeda, M. (1983). Contribution of Opponent-Colour Channels to Brightness. In Color Vision: Physiology and Psychophysics. (eds. J.D. Mollon and L.T. Sharpe). Academic Press, London.

Young, T. (1802). On the Theory of Light and Colours. Phil. Trans. R. Soc. Lond., 92, 12-48.

Opponent-Colours Theory in the Light of Physiological Findings

D. Jameson

Department of Psychology, University of Pennsylvania, 3813-15 Walnut St., Philadelphia, Pennsylvania 19104, USA

INTRODUCTION

The idea of opponent or antagonistic processes as the basic physiological substrate of our ability to perceive colors was originally developed by E. Hering before the discovery of what was initially called visual purple, and long before anything at all was known about cone receptors or the cone photopigments. Hering's antagonistic mechanisms were thought by him to be neural, and to be somehow associated with opposite neurochemical processes. Moreover, although Hering's idea was formulated as a three-variable concept, and although he published quantitative studies of metameric color-mixture and was concerned with matters such as intensity invariance of color matches and the influence of macular pigment on such data, he was not concerned with quantification of the psychophysical data base for the opponent mechanisms per se.

Our own work in color and color theory, starting in the late 1940's and early 1950's, was concerned with quantitative psychophysics and the development of a quantitative theoretical model that would both encompass the experimental data and also give natural expression to the qualitative perceptual phenomena of color vision that had long been of interest to perceptual psychologists (as well as to artists through the ages). In retrospect, it is surprising that so little was known at that time about the physiological mechanisms on which color vision depends. One could state the constraints on mechanisms that were imposed by the psychophysical findings. One could be sure, for example, that there must be a triplex of cone sensitivities and that these sensitivities must be linear transforms of the color-mixture data, that the hue-coding mechanisms must be such as to account for both hue pairing and hue antagonism, and that the processing mechanisms must be dependent on spatial interactions.

Cone Sensitivities and Sensitivity Control

Although George Wald had successfully used a selective bleach technique to determine the difference spectrum of a cone pigment, iodopsin, from a mixed rod and cone retina, the first spectra for individual human and primate cones were reported only in 1964 (Brown and Wald; Marks, et al.). Not until the 1980's was there an adequate reported sample of individual cone spectra for humans and primates to provide a reasonably firm estimate of their average forms and peak absorptions as well as an estimate of the variability of such measured spectral absorptions (Dartnall, et al., 1983).

Much experimental effort had meanwhile been expended to deduce the postulated three cone sensitivities from psychophysical adaptation experiments of different sorts. The importance of color adaptation becomes obvious as soon as one looks at the calculated chromaticities of a series of reflectance samples when illuminated by a standard source for simulated daylight, on the one hand, and their changed chromaticities when illuminated by a standard source for indoor, incandescent illumination. The relative spectral energies of two such illuminants are shown in Figure 1. Figure 2 plots chromaticity lines for Munsell samples of five different constant hue designations for the two different standard sources of illumination. The change to incandescent light is sufficient to alter the chromaticity of a sample designated as blue in the Munsell color system so that it becomes identical to the chromaticity of a yellow sample that is illuminated by the daylight source. In a foveal, bipartite color-matching field and dark surround, the Munsell "blue" sample illuminated by Source A would match the Munsell "yellow" sample illuminated by Source C. This would not, of course, be the case were we to look at the whole array of samples illuminated by Source C at one time and by Source A at another. Although we all know that color constancy is only approximate, we all also know that a blue-bird would not be mistaken for a goldfinch if we brought it indoors (Jameson, 1983).When we started to work in color, the standard way of analyzing this problem of surfaces that change chromaticities with illumination but remain relatively constant in hue, was to assume the validity of the von Kries coefficient rule. Essentially, this postulate states that each spectrally selective visual mechanism alters its overall level of sensitivity in response to light stimulation in such a way as to renormalize the

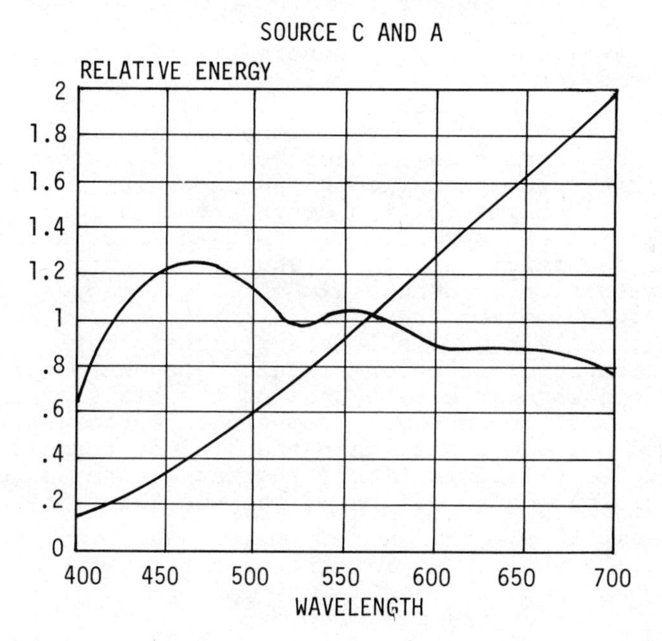

Figure 1. Spectral distributions of two standard illuminants.

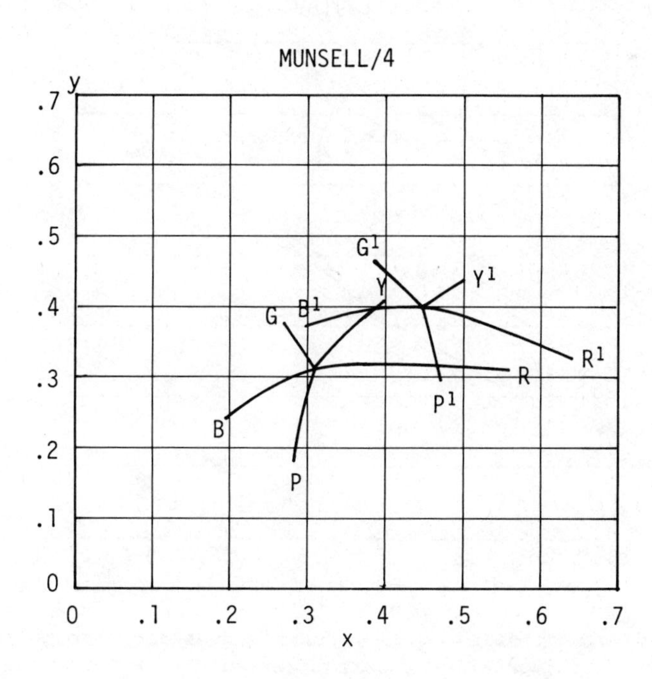

Figure 2. Chromaticities of five Munsell hues with two illuminants.

three relative sensitivities with respect to the prevailing illumination. A renormalization of the three curves used for the standard chromaticity determinations shifts the specification of the incandescent source back to the initial chromaticity of the daylight source, and the colored samples shift back toward their initial chromaticities under daylight, as shown in Figure 3. The renormalized chromaticities depend, for their precise locations, on the specific curves used as candidates for the cone sensitivities. The standardized curves used for the illustration hold no claim for legitimacy as cone sensitivities, since they are no more than computationally convenient linear transformations of standardized average color-mixture data. Were we to construct a chromaticity system on the basis of three cone sensitivities derived from the recent measures of human and primate cones, then the kind of normalization with change in illumination illustrated here would be a more legitimate application of the von Kries principle.

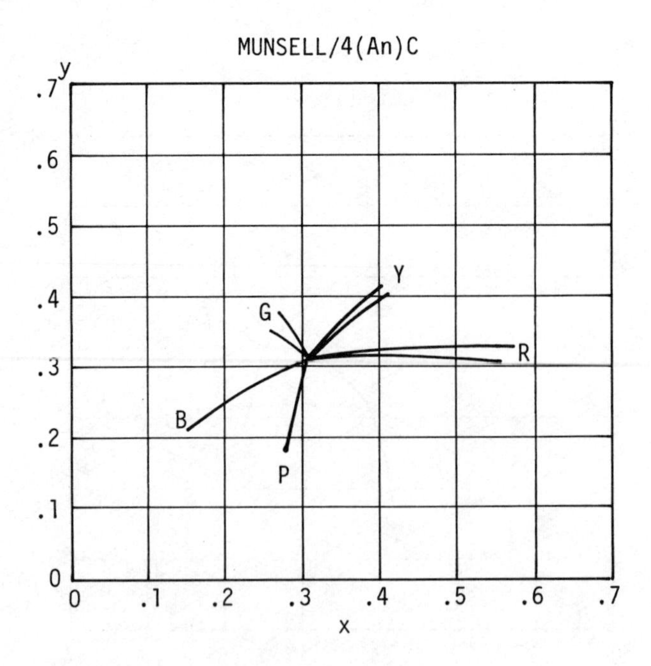

Figure 3. Chromaticities of five Munsell hues for Source C, and for Source A with von Kries renormalization.

There would, however, be little point in illustrating such a treatment, because, as we learned in the 1950's, adjustment of receptor sensitivities is only a part of what happens in the processing of color information that is relevant to the whole issue of approximate color constancy and changing illumination. The von Kries Law, or any other computational name for such a receptor amplitude normalization procedure, is inadequate. Such computational approaches require that all reflectances change appearance in the same way provided that their relative spectral reflectances are identical to each other. The overall level of reflectance must be of no consequence. But this requirement is falsified by the dependence on level known qualitatively as the Helson-Judd Effect and shown quantitatively many years ago by our own asymmetric-adaptation, color- matching data (Hurvich and Jameson, 1958). As the simplest example, spectrally neutral reflectance samples will take on the hue of the illuminant at high reflectance levels, they will remain constant and neutral at intermediate reflectances, and they will take on a hue complementary to that of the illuminant at low reflectances. A multiplicative von Kries normalization, or a Retinex ratio computation (Land, 1983), is inadequate to deal with these systematic effects. To deal with them, we must consider both receptoral sensitivity adjustments and opponent processing at the postreceptoral neural level. Let us see, then what a schematic model of the color vision mechanism must include, in its most economical form.

Opponent Colors Model

The next figure shows such a schematic model. Receptors of all three kinds are related to each of the opponent processing neural mechanisms. We can think of these mechanisms as ultimately coding redness and greenness in one of the three variable systems, blueness and yellowness in another, and achromatic neutrals in a third. The left-hand side of the diagram in Figure 4 that shows multiple connections from receptors to postreceptoral mechanisms might be thought of as analogous to receptive field center mechanisms. The lateral connections from and to the postreceptoral mechanisms schematized on the right-hand side of the diagram are analogous to receptive field opponent surround mechanisms. This kind of schematic model, based as it is on requirements of psychophysical and perceptual data for intact organisms, does not of course reveal the detailed stages of cellular processing through different layers of the retina, lateral geniculate body and the various projection areas of visual cortex.

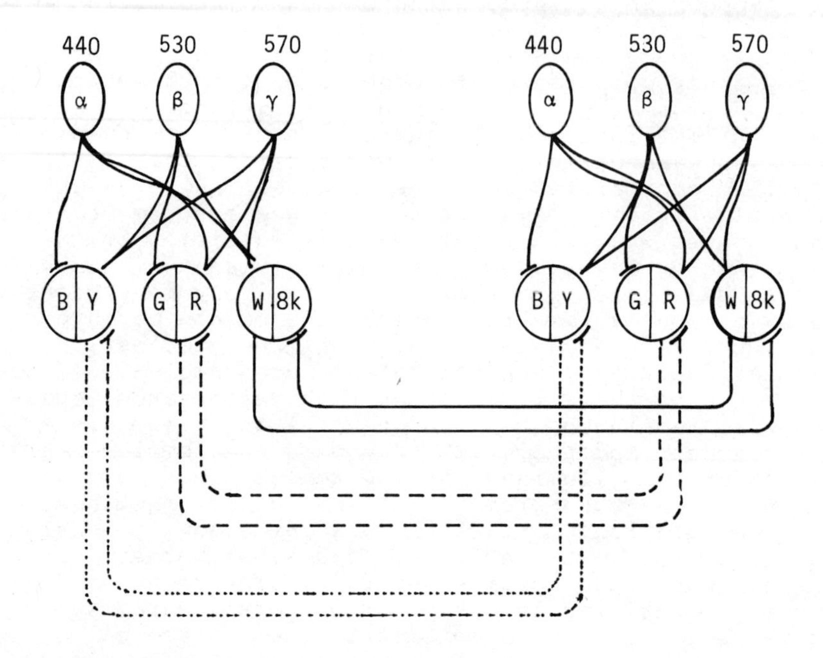

Figure 4. Schematic opponent colors model.

Conceptually, the basic requirements of the opponent-process theory are quite simple. Light absorption occurs with a tripartite spectral separation in three cone types with maximal responsiveness near 440nm, 530nm and 560nm; excitatory and inhibitory processes provide for spectral comparisons by appropriate addition and subtraction of the separate cone signals, and excitatory and inhibitory processes also provide for spatial comparisons after the spectral processing has occurred. There is summation across receptive field centers and also within receptive field surrounds, and subtraction spatially between center and surround activities. Adjustments of the system to changing ambient levels of retinal stimulation are made by compensatory amplitude adjustments within each spectrally selective set of cones and also by what appear to be set-point changes of the equilibrium level between excitatory and inhibitory processes at the postreceptoral stages (Jameson and Hurvich, 1964, 1972; Jameson et al, 1979).

With respect to the spectral processing that transforms unimodal receptor processes to bimodal opponent processes, the psychophysical measures are the data of hue cancellation, and they are supplemented by data of hue and saturation scaling (Hurvich, 1981). For cellular correlates of the red/green hue responses, we would anticipate opponent center mechanisms, with responses of common sign starting from the shortwave extreme of the spectrum up to the unique blue at about 475nm, and again starting from the unique yellow at about 580nm out to the longwave end of the spectrum, with opposite responses throughout the spectral region between 475 and 580nm. The cells typically dubbed "red/green opponent" by neurophysiologists (Livingstone and Hubel, 1984) often show no shortwave "red" lobe, or, when the shortwave lobe is recorded, the cell might be labelled "trichromatic" by the physiologist (De Monasterio et al, 1975). The nomenclature used by many neurophysiologists differs from our own and that of most psychophysicists. The former tend to assign hue names to the three cone types, and then to carry these names on to the cells in terms of the deduced receptor origins of the cellular responses. Our own strong preference, in the interest of clarity, has been to limit hue names to the experiential qualities, and by extension, only to their candidate physiological correlates. With respect to the anticipated spectral loci at which cells showing center "red/green" opponency would reveal response reversal, the wavelengths specified above are based on average psychophysical data for intact human observers with normal color vision. They are, moreover, anticipated loci for such observers only when measured under carefully controlled experimental conditions designed to guarantee that the visual system remains in a chromatically neutral state of adaptation. In the reports of physiological experiments in which single cell recordings of spectral responses are made, it seems to be assumed that the adaptation state of the animal is "neutral" unless the responses are deliberately recorded for lights superimposed on a strongly chromatic background. Prior exposure history or interstimulus intervals are seldom mentioned. "White light" background illumination used to maintain a photopic state in such studies might be light from any unfiltered broadband source, and the spectral distributions of two such "white" light sources could differ as much as the two standard sources shown in Figure 1 from one laboratory to the next. Hence the range of wavelengths in which response reversal seems to occur in different cells is neither easily interpretable nor easily compared with the human psychophysical data. It is not always obviously clear that cells labelled "red/green" might not include some that would more closely correspond to the blue/yellow opponent system, as defined psychophysically. Cells corresponding to this opponent hue system would be anticipated to have one mode of response from the shortwave spectral extreme as far into the spectrum as 500 to 510nm, and an opposite response mode beyond this wavelength region to the longwave end of the spectrum. There are occasional reports of cells that the physiologists label "blue/yellow", often it is reported that such cells are not seen, and sometimes the habit of labelling cells by referral back to hue-named cones causes doubts about the very existence of such a category. If the cone with maximal sensitivity near 560nm (a spectral region that typically looks greenish-yellow to my neutrally adapted eye) has been christened "red", it is not hard to see that such a naming convention could interfere with classification of hue opponency in different cell systems.

Whether the primate visual system has only a sparsity of cellular mechanisms that might be correlates for the blue/yellow opponent system of human color vision, or whether there is some other difficulty related to sampling or criteria for identifying such cells seems, at present, to be an unanswered and perhaps not immediately answerable question. Certainly there has been a change between the earliest findings on the relative numbers of spectrally selective cells in the primary visual cortex and studies a decade or so later, for example, by Michael (1978a-c,1979,1981), and most recently by Livingstone and Hubel (1984). Also there still seems to be disagreement among neurophysiologists about the proportions of cells specifically involved in color processing in the extra-striate visual area V4 (Schein et al, 1982; De Monasterio and Schein, 1982; Zeki, 1983a,b). We hesitate, therefore, to credit any particular significance to relative numbers of cells of different types or to their numbers in particular visual areas of the primate brain.

We would like particularly to have more information about systems that might code blue and yellow because, in psychophysical experiments, this system seems to have some properties that differ from the red/green one. All tests of red/green linearity seem to yield positive outcomes, whether they are tests of cancellation invariance as a function of intensity, additivity tests, or simply goodness-of-fit tests for linear models. When it is the blue/yellow system that is examined, all such tests yield a variety of outcomes for different individual observers. Some individuals behave linearly, many do not, and some show grossly nonlinear behavior (Werner and Wooten, 1979). Our own cancellation data are reasonably well fit by linear transformations of receptor sensitivities with deviations falling within the error of the psychophysical measures, and this is true both for the red/green and the blue/yellow measures. These psychophysical issues are dealt with by Hurvich, in this symposium. They do, however, point up the need for more information about possible physiological correlates of the blue/yellow opponent hue mechanism.

With respect to the black/white, achromatic system, the physiological mechanisms of excitation and inhibition in spatially organized receptive fields was known even before Svaetichin's (1956) first demonstration of spectral opponency in graded polarization responses in the fish retina. What becomes important now, for the achromatic system as well as for the chromatic ones, is the organization and dimensionality of receptive fields of cells related to different regions of the retina, and also the significance for perceptual information of many different parallel systems and different stages of hierarchical processing.

Receptive Fields: Models and Questions

It is possible to develop a color vision model that conceptualizes sensitivity control at the three-variable receptor level and lateral and temporal interactions at the three-variable opponent-process level and thereby to account for a variety of color and brightness constancy, contrast and after-effect data (Jameson and Hurvich, 1961a,b, 1964, 1972). Only by the use of empirically determined constants in the spatial interaction equations, however, are the different effects that depend on retinal locus and retinal image dimensions and layout incorporated in the account.

Clearly, the grain of the receptor mosaic and, more importantly, the grain of the postreceptoral processing systems need to be known or postulated. It has become conventional in psychophysical or computational models (Marr, 1982) to use a standard representation of circularly symmetric, center-surround receptive fields that is described mathematically as a difference of two Gaussians (DOG). In the computer printout of Figure 5, the center-surround proportions (widths about 1:9 and excitation to inhibition about 1.6:1) are those originally proposed by Bekesy (1968), and we have used units of this sort to model a variety of contrast and assimilation effects (Jameson and Hurvich, 1975, 1983). The unit shown in Figure 5 has a small center diameter (0.05deg), and it is assumed that field centers increase monotonically with retinal eccentricity, an assumption consistent with anatomy, physiology and psychophysics. It is also assumed that there must be a distribution of receptive field sizes topographically related to each given region of the retina. This assumption is consistent with analyses of spatial vision in terms of a Fourier-like analyzer mechanism made up of a number of so-called channels tuned to different spatial frequencies. There is at least suggestive evidence from primate visual cortex that can be interpreted in this way (De Valois et al, 1982a,b). For three different sizes of center-surround receptive fields of the proportions shown in Figure 5, unit responses to spot stimulation of increasing radius are calculated to have the forms shown in Figure 6. This plot is analogous to unit recording data in which spot size is progressively increased to determine the diameter of a receptive field center. It would be useful to have such plots for a sufficiently large sample of single cells to make an estimate of the distribution of receptive field sizes and shapes at least for concentric and simple cells, and also for cells that are dual opponent as well as those that have only one mode of center response.

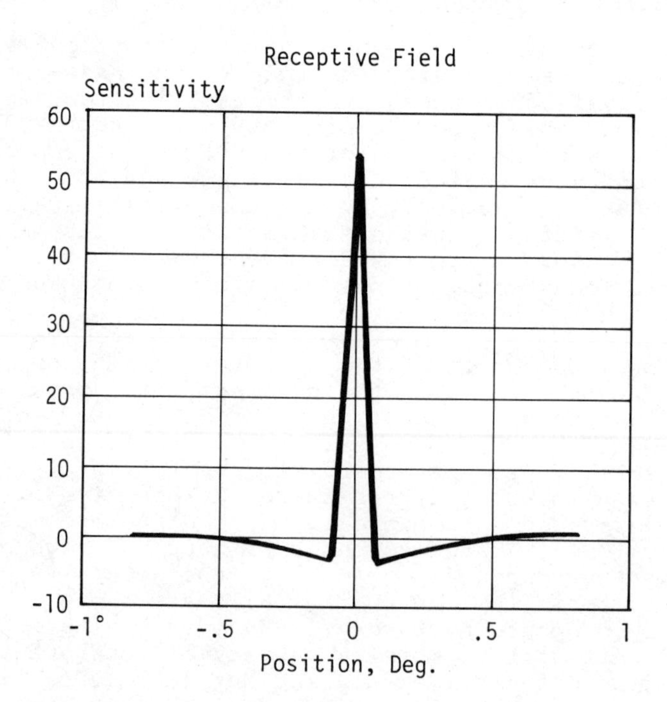

Figure 5. Receptive field sensitivity profile (DOG).

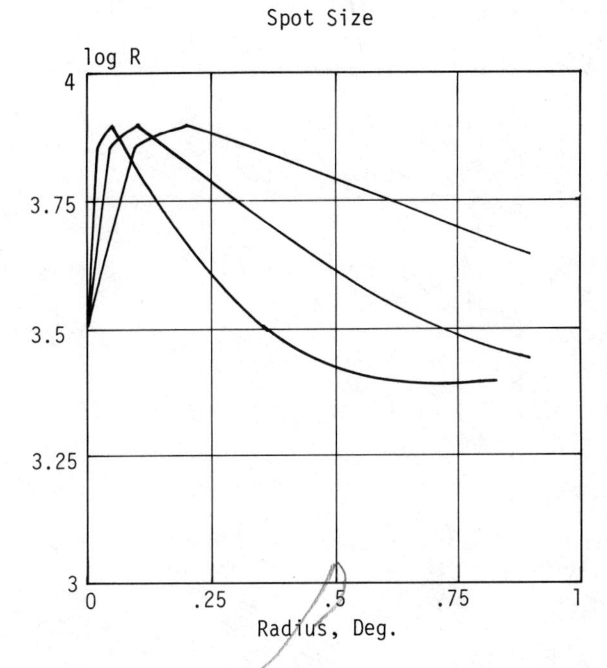

Figure 6. Computed responses of three different single units with increasing size of uniform stimulus.

The importance of receptive field grain becomes quite clear when we consider a phenomenon such as border contrast. A simple gray scale suffices as an example. If the steps of the scale are butted, one against the other, we see the familiar scallop or fluted effect. Whether the brightness-difference enhancement at each border is seen as a bend, a more rounded shading, or something approaching a blunted saw-tooth depends on the distance at which the gray scale is viewed, and also on whether we are concentrating on the adjacent gray steps just in the fixation area or whether we are broadening our range of attention focus to weight more heavily the more peripherally imaged steps as well.

Figure 7 shows, above, a luminance profile of such a butted gray scale, and, sequentially below, computed brightness profiles based on one-dimensional convolution of DOG sensitivity functions as in Figure 5 but with unit, and two doublings in width, respectively. The computed effect is the same if one assumes a fixed viewing distance and three different receptive field diameters, or a single receptive field diameter and three different viewing distances.

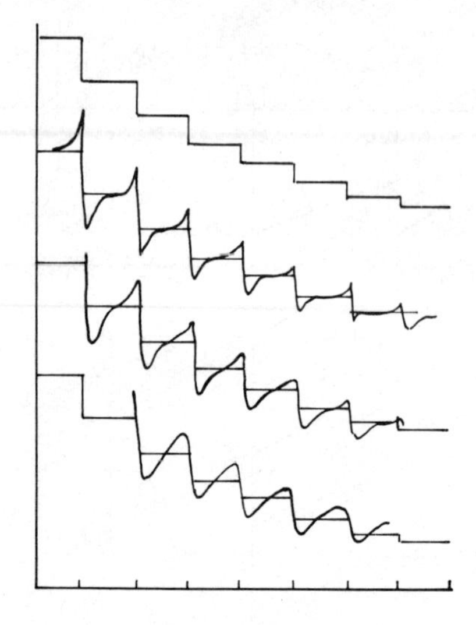

Figure 7. Luminance profile of gray scale and computed contrast effects for increasing size ratio of receptive field to retinal image.

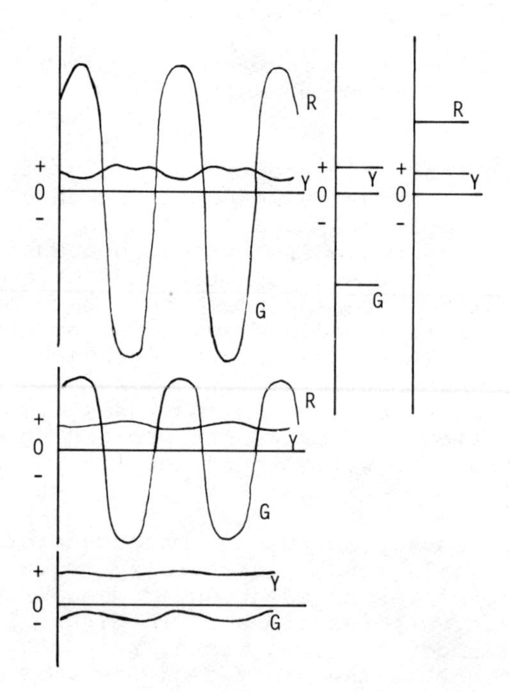

Figure 8. Computed hue responses for two reflectances. Right: fo] two uniform fields; left: for alternating stripes and increasing size ratio of receptive field to retinal image.

Let us now turn to an example in which the grain of image sizes relative to receptive field sizes determines whether hue differences between a yellowish green and a yellowish red are accentuated as color contrast effects or nearly averaged out as color assimilation effects. The original pattern is a layout of just three reflectances, printers' inks that, seen in isolation, are a red, a green and a blue, respectively. Our computational illustration will be concerned only with the red and green pair. These reflectances, together with a standard illuminant (Source A), are evaluated separately by each of the three cone sensitivities, and the resulting three values for each of the two reflectances are entered in the equations that relate the cone "signals" to the opponent hue systems (Hurvich, 1981). The computed values for redness and yellowness, and for greenness and yellowness are shown on the right hand side of Figure 8. Positive and negative signs follow the usual convention in opponent colors theory (redness +, greenness -, yellowness +, blueness -). These computed hue amounts do not take into account the spatial layout and dimensions of the two reflectance surfaces in the pattern. This is done for the alternating stripes with the results shown on the left hand side of the figure, again computed by one dimensional convolution with receptive field profiles as in Figure 5, but in this instance, dual opponent ones. Again, the computations are made for three different scales, interpretable either as different receptive field widths at a fixed viewing distance, or as a single size of receptive field and three viewing distances. The effect varies from hue contrast where the stripes in the retinal image are wide relative to the receptive field center diameter to hue assimilation (a kind of spatial averaging) where the stripes are narrow relative to the receptive field diameter. One can readily verify the qualitative correctness of such a computational approach by moving such a pattern closer and farther away. Or, it can be verified in a more interesting way by studying a Seurat or Signac pointillist or divisionist painting at very close range and then backing away from it. Depending on the painting, there will be a best distance for the viewer. For me, this is a fairly close distance at which the pigment elements are separately resolvable in the fixation area but blend together into more continuous surface colors in the parts of the picture at which I am not directly looking. As the fixated region changes from moment to moment, the painting takes on a particular quality of visual liveliness.

In modeling color effects, we have started by using proportions and sizes for dual-opponent concentric receptive field profiles that are the same as for the achromatic system, and that do not differ for the red/ green and yellow/blue processing mechanisms. Also, we have restricted the computational scheme to circularly symmetric receptive fields. The latter restriction is based on the assumption that in the visual physiology such units enter in size graded fashion into the hierarchical processing that forms simple cells tuned to line width and orientation, and leads progressively to other cells with more complex response properties.

Our starting point for the three opponent systems, which implies that the achromatic and chromatic systems differ only in terms of spectral opponency, is a conscious oversimplification that we recognize as unrealistic. Study of the temporal properties of the systems makes it clear that temporal discriminations that do not involve fluctuations in hue exhibit band-pass temporal frequency characteristics, whereas temporal discriminations of equiluminous heterochromatic stimuli have low-pass characteristics that are similar for red/green and for yellow/blue (Varner et al, 1984). This difference in temporal processing is the basis for the traditional and reliable use of heterochromatic flicker photometry.

The spatial characteristics of the three opponent systems have only recently begun to be separately examined and analyzed, but there is much current interest in this issue (De Valois et al, 1984; Knoblauch et al, 1984; Mullen, 1984). The work with Knoblauch in our own laboratory is concerned with the contribution of chromatic differences to threshold measures of sinewave contrast sensitivity, and with the relation of these measures to suprathreshold hue contrast effects in squarewave patterns. The stimulus variables are chosen so as to measure separately the effects for the red/green and yellow/blue chromatic systems. Thus for the threshold measures, in one experimental situation the uniform light component is a unique yellow, 580nm, and the sinewave component is the same wavelength for pure luminance contrast, a shorter wavelength to add greenness, or a longer one to add redness. In the second situation the uniform light is a unique green at 505nm, and the sinewave component is the same for pure luminance contrast, a shorter wavelength to add blueness, or a longer one to add yellowness. For the suprathreshold hue contrast measures, the criterial measure was a wavelength setting for unique yellow in alternate bars of the squarewave pattern, in the presence in the adjacent bars of the same series of wavelengths used in the threshold experiments, or else a wavelength setting for unique green, again in the presence of the same series of wavelengths used in the threshold experiments. In the suprathreshold experiments luminance was held constant, so the wavelength shifts required to meet the unique hue criteria were measures of pure hue contrast.

The data for one observer are shown in Figure 9. The lower two curves in the upper graph are for a spatial frequency of 0.26 c/deg and the higher two curves are for 1.3 c/deg. The increase in sensitivity between the two pairs of curves is attributable to the luminance system. The increment in contrast sensitivity attributable to blueness or yellowness on the one hand or redness or greenness on the other varies with the wavelength pairings, but it is the same linear increment at both spatial frequencies. This is shown on the graphs by the solid points that plot the sum of the luminance sensitivity at 1.3 c/deg and the chromatic increment in sensitivity determined at 0.26 c/deg. The agreement with the experimental data is good enough to conclude that chromatic and luminance contrast sensitivities simply add linearly at threshold and also that, although luminance contrast sensitivity increases between 0.26 and 1.3 c/deg, both yellow/blue and red/green spatial contrast sensitivities are approximately constant in this spatial frequency region. This behavior is consistent with spatial low-pass characteristics for both the yellow/blue and red/green systems, and band-pass characteristics for the achromatic luminance system. If the threshold measures can be used to predict suprathreshold characteristics, these measures suggest that the wavelength shift indices of hue contrast should be identical at both spatial frequencies (in this experiment 0.27 c/deg and 1.6 c/deg), but again dependent on the wavelength of the alternate bars in the squarewave grating. This is the result shown in the lower graph of Figure 9. The data for the two spatial frequencies plot nearly on top of each other for both the yellow/blue and red/green hue contrast measures.

Results of this sort, which are consistent with reports from other laboratories (Mullen, 1984), suggest that receptive field modeling should assume larger field centers for dual-opponent sensitivity profiles than for achromatic, spatially opponent ones. Whether the proportions of the excitatory and inhibitory regions should differ as well is not answered by such data. Although it would not be surprising to find that field centers differ in size between the two dual-opponent systems, more evidence is needed to make specific proposals about these potential dimensional differences.

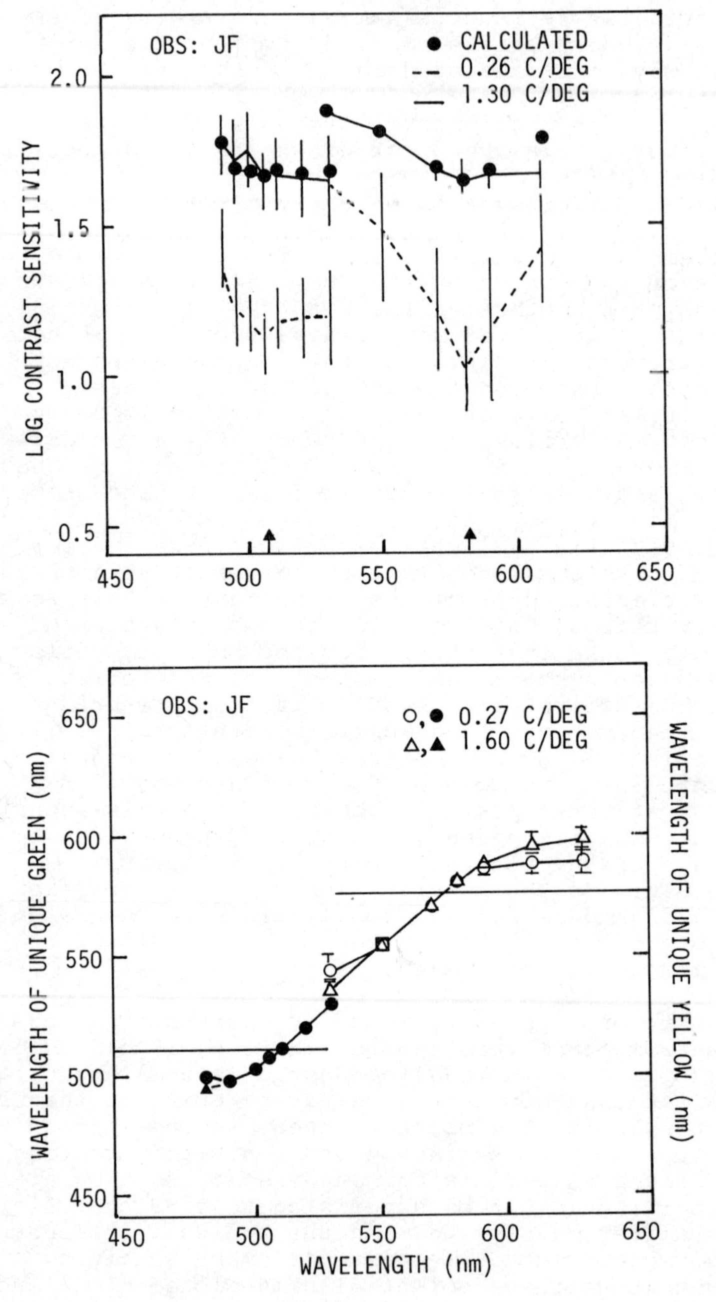

Figure 9. Spatial contrast sensitivity as functions of wavelength difference (above), and wavelengths for unique green and unique yellow as functions of spatially adjacent wavelengths (below).

Visual Physiology and Color Coding

Obviously, much is still to be learned in order to model many aspects of the color vision system in physiologically realistic terms. It also seems obvious that much is still to be learned about the visual physiology itself: the parallel systems within specific visual areas and in different visual areas, the communication or interactions among them, and their specialized functions in physiological color processing. Ultimately, one would like to see a general systems analysis of the sort that is being attempted for the oculomotor system. But there is a fundamental difference that we must be resigned to. In the oculomotor system we can at least specify a retinal input, a layout of retinal locations, a motor output, an eye/head/body movement, and a result, a changed retinal layout. An intentional motive and a conscious readout might be assumed or inferred, but need not be physiologically specified to describe the workings of the input/output system.

For color vision, we have no such objectively measureable endpoint or output that would enable us to decide whether the dual-opponent cells that are found in "blobs" in the visual cortex should be hue coded or simply assigned to some stage in one or another processing system. Nor do we have a criterion for deciding how the cells in V4 with their different spectral response properties enter into either the coding or the processing of the hues, saturations and lightnesses that we humans know as color vision. Certainly, cells that fail to be at all responsive to changed stimulation in the immediate surround of their centered stimulation would seem to have little to do with what we all see. Insulated private lines from retinal receptors to visual brain centers were once thought of as even likely for the foveal region of the retina, but would find few if any supporters among theorists today.

In conclusion, let me say in relation to the title of this contribution that in the light of physiology, opponent colors theory is alive and well. The kinds of mechanisms that the theory postulates, spectral opponency and spatial opponency, are physiologically in evidence. The particular spectral response characteristics that would identify particular types of cells as correlates of the three separate, white/black, yellow/blue and red/green mechanisms that we deduce from visual psychophysics are less satisfactorily established in the visual physiology. This is also the case for the spatial response properties of cells that seem to be involved in hue and achromatic spatial discriminations and interactions. Further theoretical developments to handle systematically the subtle and not so subtle aspects of color vision that enter into our perceived and internally represented worlds will depend on further advances in both visual psychophysics and visual physiology, and will proceed on surer footing with continued and more intensive interaction between the two disciplines.

References

Bekesy, G. von. (1968). Mach- and Hering-Type Lateral Inhibition in Vision. Vision Res., 8, 1483-1499.

Brown, P.K. and Wald, G. (1964). Visual Pigments in Single Rods and Cones of the Human Retina. Science, 144, 45-52.

Dartnall, H.J.A., Bowmaker, J.K. and Mollon, J.D. (1983). Microspectrophotometry of Human Photoreceptors. In Colour Vision: Physiology and Psychophysics. (eds. J.D. Mollen and L.T. Sharpe). Academic Press, London.

De Monasterio, F.M., Gouras, P. and Tolhurst, D.J. (1975) Trichromatic Colour Opponency in Ganglion Cells of the Rhesus Monkey Retina. J. Neurophysiol., 251, 197-216.

De Monasterio, F.M. and Schein, S.J. (1982). Spectral Bandwidths of Color-Opponent Cells of Geniculocortical Pathway of Macaque Monkeys. J. Neurophysiol., 47, 214-224.

DeValois, K.K., Webster, M. and Switkes, E. (1984). Orientation Selectivity for Luminance and Color Patterns. (A). Invest. Ophthal. and Vis. Sci. (Suppl.), 25, 232.

De Valois, R.L., Yund, E.W. and Hepler, N. (1982a). The Orientation and Direction Selectivity of Cells in Macaque Visual Cortex. Vision Res., 22, 531-544.

De Valois, R.L., Albrecht, D.G. and Thorell, L.G. (1982b). Spatial Frequency Selectivity of Cells in Macaque Visual Cortex. Vision Res., 22, 545-559.

Hurvich, L.M. (1981). Color Vision. Sinauer, Sunderland, Mass.

Hurvich, L.M. and Jameson, D. (1958). Further Development of a Quantified Opponent-Colours Theory. In Visual Problems of Colour II. Her Majesty's Stationery Office, London.

Jameson, D. (1983). Some Misunderstandings about Color Perception, Color Mixture and Color Measurement. Leonardo, 16, 41-42.

Jameson, D. and Hurvich, L.M. (1961a). Opponent Chromatic Induction: Experimental Evaluation and Theoretical Account. J. Opt. Soc. Am., 51, 46-53.

Jameson, D. and Hurvich, L.M. (1961b). Opponent-Colors

Theory and Physiological Mechanisms. In The Visual System: Neurophysiology and Psychophysics. (eds. R. Jung and H. Kornhuber). Springer, Berlin.

Jameson, D. and Hurvich, L.M. (1964). Theory of Brightness and Color Contrast in Human Vision. Vision Res., 4, 135-154.

Jameson, D. and Hurvich, L.M. (1972). Color Adaptation: Sensitivity, Contrast, After-Images. In Handbook of Sensory Physiology, 7/4, Visual Psychophysics. (eds. D. Jameson and L.M. Hurvich). Springer, Berlin.

Jameson, D. Hurvich, L.M. (1975). From Contrast to Assimilation: In Art and in the Eye. Leonardo, 8, 125-131.

Jameson, D. and Hurvich, L.M. (1983). Color Processing Beyond Mondrian. (A). J. Opt. Soc. Am, 73, 1885.

Jameson, D., Hurvich, L.M. and Varner, F.D. (1979). Receptoral and Postreceptoral Visual Processes in Recovery from Chromatic Adaptation. Proc. Natl. Acad. Sci. USA, 76, 3034-3038.

Knoblauch, K., Jameson, D., Hurvich, L.M. and Geller, A. (1984). Chromatic Factors in Threshold and Suprathreshold Spatial Vision. (A). Invest. Ophthal. and Vis. Sci. (Suppl.), 25, 232.

Land, E.H. (1983). Recent Advances in Retinex Theory and Some Implications for Cortical Computations: Color Vision and the Natural Image. Proc. Natl. Acad. Sci. USA, 80, 5163-5169.

Livingstone, M.S. and Hubel, D.H. (1984). Anatomy and Physiology of a Color System in the Primate Visual Cortex. J. Neurosci., 4, 309-356.

Marks, W.B., Dobelle, W.H. and MacNichol, E.F. (1964). Visual Pigments of Single Primate Cones. Science, 143, 1181-1183.

Marr, D. (1982). Vision. Freeman, San Francisco.

Michael, C.R. (1978a). Color Vision Mechanisms in Monkey Striate Cortex: Dual-Opponent Cells with Concentric Receptive Fields. J. Neurophysiol., 41, 572-588.

Michael, C.R. (1978b). Color Vision Mechanisms in Monkey Striate Cortex: Simple Cells with Dual Opponent-Color Receptive Fields. J. Neurophysiol., 41, 1233-1249.

Michael, C.R. (1978c). Color-Sensitive Complex Cells in Monkey Striate Cortex. J. Neurophysiol., 41, 1250-1266.

Michael, C.R. (1979). Color-Sensitive Hypercomplex Cells in Monkey Striate Cortex. J. Neurophysiol., 42, 726-744.

Michael, C.R. (1981). Columnar Organization of Color Cells in Monkey's Striate Cortex. J. Neurophysiol., 46, 587-604.

Mullen, K.T. (1984). Human Contrast Sensitivity to Blue/Yellow and Red/Green Chromatic Gratings. (A). Invest. Ophthal. and Vis. Sci. (Suppl.), 25, 232.

Schein, S.J., Marrocco, R.T. and De Monasterio, F.M. (1982). Is There a High Concentration of Color-Selective Cells in Area V4 of Monkey Visual Cortex? J. Neurophysiol., 47, 193-213.

Svaetichin, G. (1956). Spectral Response Curves from Single Cones. Acta Physiol. Scand. 39 (Suppl.), 17-46.

Varner, D., Jameson, D. and Hurvich, L.M. (1984). Temporal Sensitivities Related to Color Theory. J. Opt. Soc. Am. A, 1, 474-481.

Werner, J.S. and Wooten, B.R. (1979). Opponent Chromatic Mechanisms: Relation to Photopigments and Hue Naming. J. Opt. Soc. Am. 69, 422-434.

Zeki, S. (1983a). Colour Coding in the Cerebral Cortex: The Reaction of Cells in Monkey Visual Cortex to Wavelengths and Colours. Neuroscience, 9, 741-765.

Zeki, S. (1983b). Colour Coding in the Cerebral Cortex: The Responses of Wavelength-Selective and Colour-Coded Cells in Monkey Visual Cortex to Changes in Wavelength Composition. Neuroscience, 9, 867-781.

Receptoral Constraints on Colour Appearance

D.I.A. MacLeod

Psychology Department, University of California at San Diego, La Jolla, California 92093, USA

Goethe wrote: "All theory is gray," and his complaint is particularly applicable to studies of the earliest processes in color vision - the events at the receptor level. The phenomena of color vision that impress us most are seldom traceable to receptor behaviour. But still it is important to be clear about the ways receptors constrain sensation and the ways they don't. Most obviously, the receptors are the sole source of the information on which color vision depends. Radiation that fails to stimulate the receptors can't be seen; and stimuli that are equal to one another in their effects on the receptors have to appear identical if viewed under similar conditions. We do tend to forget this, because we generally know the stimuli only through our own visual reactions to them; but, when we neglect the physical diversity of color stimuli in this way we can be seriously misled. For instance, we may fail to appreciate the basic point that surfaces that are visually indistinguishable under one source of illumination will in general appear different from one another when the illuminant is changed. This type of deviation from perfect color constancy (which Dr. Land has also noted in his presentation) has a lot of aesthetic and commercial importance, and is an inevitable consequence of what Rushton (1971) called the Principle of Univariance, that the signal from a receptor depends only on its state of excitation, and not on the wavelengths of the exciting stimulus. Since any account of color appearance, under constant or varying conditions of illumination, must start by taking account of how much the receptors are excited by different wavelengths, I begin by very briefly reviewing our knowledge about that, then go on to a closely related issue: the physiological basis of trichromacy. Trichromacy, the fact that three independent adjustments are necessary and sufficient for matching any color, is thought to be a consequence of the existence of only 3 cone types whose excitations have to be equated for a visual match. Although there can be no argument with the proposition that 3 classes of cone exist in normal retinas, I am not satisfied with this account of trichromacy. I will show a clear counter-example leading to the conclusion that trichromacy has its origin in a trivariance of neural organization. Finally, I describe briefly some experiments that bear on the contribution that receptor adaptation makes to color constancy.

First, then, as to the spectral characteristics of the receptors, it is enough for our present purposes to recognize that they are pretty much what Konig believed them to be nearly a hundred years ago. There are 3 cone types, with distinct but extensively overlapping spectral sensitivities. These broad spectra carry some penalty for color discrimination with natural broadband stimuli, but they are essential for retaining good visual sensitivity since if individual cones absorbed light in only a narrow spectral band, all of the incident quanta belonging to other spectral bands would inevitably be wasted.

Each cone generates a signal depending on its own excitation, and roughly speaking we can associate intensity with the sum of the three cone excitations, and chromaticity with their ratios. The receptoral basis of color vision is best appreciated graphically by using Cartesian coordinates to represent the excitations of each of the three cone types by a given stimulus as in Figure 1. The letters R, G and B are used here to index the axes corresponding to the 3 cone types; these of course stand for red, green and blue which are, roughly speaking, the parts of the spectrum that most selectively excite each cone type. To represent chromaticity independently of luminance, it's convenient to adopt a plane from this space as a 2 dimensional chromaticity diagram, for example, the unit plane R+G+B=1. This is the relatively clear Maxwell triangle, not the physiologically arbitrary CIE projection of it, which has held back insight into the visual process to such an extent that William Rushton was fond of saying that "it seems to have been designed to intimidate the young and bewilder the old." Still it is not quite the sort of diagram one would like to have for representing the relative cone excitations independently of luminance. There is evidence (Eisner and MacLeod, 1980) that the B cones make no detectable contribution to luminance as usually defined, so that in the 3D cone excitation space the constant luminance plane is not the oblique one, R+G+B=1 but the vertical one in Figure 1, R+G=1. If we adopt this plane as our chromaticity diagram (Luther, 1927), the excitations of all the cones for equiluminous stimuli can be read from linear orthogonal coordinates (MacLeod and Boynton, 1979). In Figure 2 the horizontal coordinate, r, shows R cone excitation, and if read from right to left, it shows G cone excitation. B cone excitation is the vertical axis, and the spectrum locus extends from violet at the top, down to green, through yellow to red on the right.

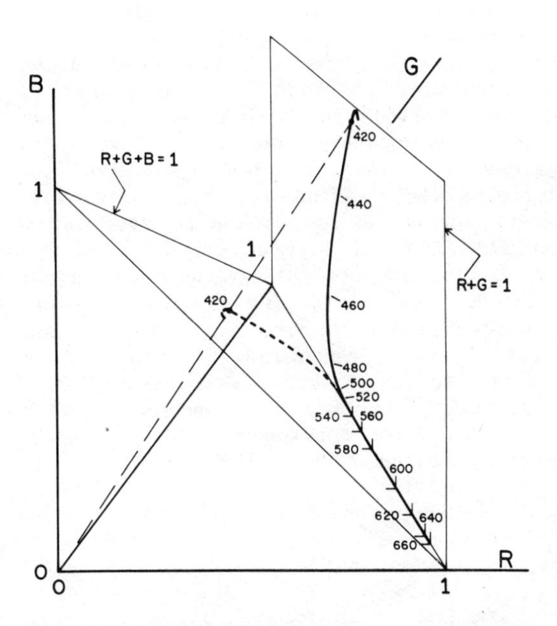

Fig. 1. Three dimensional cone excitation space, showing planes corresponding to triangular (R+G+B = 1) and rectangular (R+G=1) chromaticity diagrams.

In this cone excitation diagram, the spectrum locus never gets into a corner - there are no pure stimuli because of the spectral overlap of the receptor sensitivities. But if we were to normalize to white, the ratio of R cone to G cone excitation reaches a maximum of about 18 to 1 at 700 nm; the reciprocal ratio is greatest below 500 nm, but never gets higher than 2:1. That maximum is actually in the blue, around 465 nm. B cone excitation is rather slight for a typical white, and is more than 50 times greater in the violet. This very strong peak in the violet occurs because the spectral separation between the B cones and the other two is relatively large with maximum values of roughly 440, 540 and 560 nm. This arrangement gives the B cones a lot of leverage for color discrimination, but puts them at a disadvantage for catching quanta. Why is this better than equal spacing? It has been pointed out that chromatic aberration introduces a tradeoff between the demands of color discrimination (which is aided by spectral separation) and spatial resolution, which is hindered by it, most especially in the short wavelength spectral range. Bob Boynton (1980) has shown that spatial resolution is improved by the solution that the visual system seems to have adopted, of having the R and G cone spectral sensitivities closely spaced at long wavelengths and entrusting the task of spatial resolution entirely to them. This dependence of spatial resolution on the R and G cones is assured by excluding the B cones from the luminance system, inasmuch as it is only luminance variations that are analyzed with high spatial precision by the visual system (Granger and Heurtley, 1973). A fly in the ointment here is that physiology has not exposed the hypothetical luminance system, devoid of any B cone input, which psychophysicists like to postulate; but psychophysics does require it somehow, somewhere, and we must just wait for physiologists to find it. An incidental advantage of this arrangement whereby B cones have no role in spatial resolution, is that it allows a great economy in their numbers. David Williams, Mary Hayhoe and I (1981) got psychophysical evidence for this by mapping B cone sensitivity against long wavelength backgrounds. This revealed quite large variations in sensitivity on a very small spatial scale, the variation amounting to as much as a factor of 4 in 6 minutes of arc. We presented evidence that the peaks of sensitivity are just the regions in the visual field where B cones exist, and the insensitive valleys are the large gaps between them, filled with R and G cones.

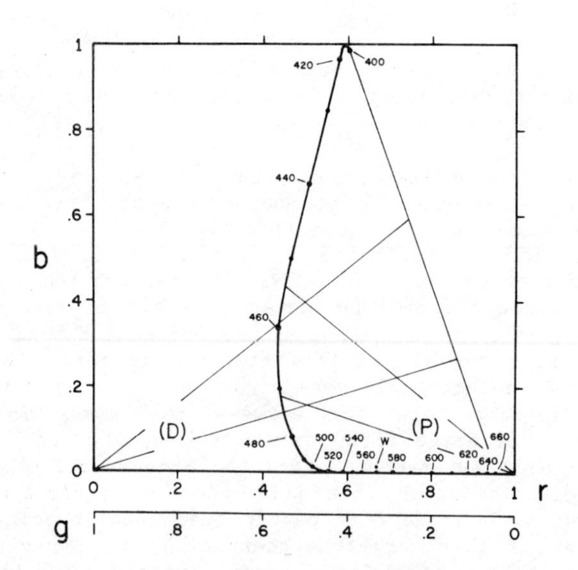

Fig. 2. Cone excitation chromaticity diagram (equivalent to the R and G =1 plane i Fig. 1). P and D indicate the confusion lines of protanopes and denteranopes, and the point W near the horizontal axis is the chromaticity of an (equal energy) white.

It is useful to consider further the fact that colors that have flattish spectra ("whites") do look subjectively neutral even though they provide only a tiny fraction of the B cone excitation available from spectral light. It is probably very useful for the visual system to have the balance between blueness and yellowness set up in this way, because physically pure blues and violets are rare in nature. Most natural stimuli are spectrally fairly flat, and a moderate skewing of energy toward long or short wavelengths is the most we are usually called on to respond to. So it is a good strategy to have the neutral point set to correspond to a fairly flat spectral reflectance, so that small deviatons from spectral flatness can be efficiently discriminated by allocating most of the response range of the color opponent cells to desaturated colors, even at the expense of discriminating power for saturated blues or yellows. Ed Pugh's contribution to this symposium gives clear evidence that nonlinearity of this kind is a prominent feature of color-opponent processing in the visual system.

One final point is made clear by this cone excitation diagram, a familiar one that was recognized by G. E. Muller in 1924. Redness and greenness are not simply correlated with excitation of the R and G cone types, that is with r. In the diagram, greenish yellows have the same r coordinates as reddish violets, and if we go from blue to yellow, keeping subjective redness and greenness zero, the R and G cones must undergo a very substantial excitation modulation, about 20%. The B cones are therefore strongly implicated in the production of redness at short wavelengths, although the nonmonotonic variation of r with wavelength in the violet-green spectral range is also a factor.

TRICHROMATIC MATCHING AND COLOR APPEARANCE

Now let us concentrate on the issue of how the cone photoreceptors constrain our sensations of color. To begin with, of course, it is usually supposed, and for good reason, that the cones impose the visual constraints implicit in color matching. Cones seem to produce only one type of signal whatever the light that excites them, an idea that William Rushton (1971) called the Principle of Univariance. Two lights of different spectral composition can be matched for any cone receptor by intensity adjustments alone, so that the two stimuli are indistinguishably represented at the cone output. With 3 cone types, each containing a different pigment, a trichromatic match is possible such that the two matched stimuli are equivalent to one another in their action on each of the 3 cone types. In this situation, no information distinguishing between those stimuli leaves the cones, and the match established at the cone level must be a subjective match as well, whatever the neural processes that intervene between the receptors and sensation. Although I think this intuition about the receptoral basis of color matching is perhaps the most fundamental and basic one in the whole field of color vision, I want to suggest it is at best a crude idealization and is, in some cases at least, fundamentally wrong.

Trichromatic Additivity and Homogeneity of the Cone Sensitivities

A most important aspect of trichromatic matches that supports Young's hypothesis that they are determined at the cones is the additivity of lights of different spectral composition. According to Grasmann's Third Law (and a related principle known as von Kries's Persistence Law) two matching lights of different spectral composition continue to match whatever additional light is exposed along with them. (Of course, the added light has to be the same on the two halves of the matching field.) This is a striking fact because as von Kries knew the actual apparent color of both the matched fields can be quite drastically changed by uniform adapting lights like this. The interpretation is that the receptor signals from the three cone types are selectively attenuated by variously colored background lights, but the equality of excitation between the 2 halves of the matching field continues to hold for each cone type, so the match still holds. If the matched fields were differently exciting the cones, and only became equivalent at some later neural stage, this addivitity or persistence of matches during adaptation would not generally be expected. Grassmann's Law stating that it does is therefore a cornerstone of the trichromatic theory.

Now we know that there are 3 cone types, but still I think this account of trichromacy is inadequate. One problem is posed by inevitable variations in spectral sensitivity between cones of the same class. Between individuals, we have evidence, from a factor analysis of color matching data (MacLeod and Webster, 1983) that cones of the same class vary in their wavelengths of peak absorption with a standard deviation very close to 1 nm. This is not much but it is enough to ensure that in general when a trichromatic match is being made, cones with this degree of variability will collectively distinguish between the matched lights - for instance the long wave displaced cones will be more excited by the half of the matching field that has red in it. The difference between the two matched fields as viewed by a single cone can exceed 10%, which is a difference that is in general easily detectable. So why do we accept trichromatic matches? One answer could be that cones of the same class in a single individual's retina are less variable than I have assumed. But microspectrophotometry does not support this, suggesting instead standard deviation substantially greater than 1 nm (Dartnall, Bowmaker and Mollon, 1983). Some psychophysical evidence about variability among the cones of an individual observer is available in the additivity of trichromatic matches. Nagy, Heyneman, Eisner and I (1981) failed to find any reliable deviations from additivity in normal subjects when suitably chosen colored backgrounds were added to the matching fields, an observation difficult to reconcile with the microspectrophotometric evidence since our calculations indicated that variability on the scale suggested by MSP would lead to easily measured failures of additivity, although variability with a standard deviation less than 1.5 nm would not. Despite these uncertainties the evidence for variability among cones is enough to raise serious doubt as to whether trichromatic matches are determined at the receptor level: the cones probably are distinguishing reliably between the matched fields.

Additivity Failure in Heterozggotes: Neural Trivariance

Fortunately, however, we have found perfectly clear violations of Grassmann's Laws in a particular, not uncommon group of subjects (Nagy, MacLeod, Heyneman and Eisner, 1981). For these subjects we can therefore say for certain that trichromatic matches are not established at the cones. These subjects are women who carry genes for anomalies of color vision. People with anomalous color vision make abnormal trichromatic matches because one of their pigments, either the R or the G pigment, has been replaced by an abnormal one. Lights that match for the normal pigments will generally be a mismatch for the anomalous pigment, and this pigment will dictate some different match. These cone pigments are produced by genes on the X chromosome, and the heterozggous carriers who do not manifest the anomaly presumably have one X chromosome making the normal pair of pigments and another where a mutated gene makes an anomalous pigment instead of one of the normal ones. Moreover, we know (from evidence recently reviewed by Luzzatto and Gartler, 1983) that in each cell of a mammalian carrier's body one or another of her two X chromosomes is dominant; the result is a mosaic of cells with characteristics given by her maternal or paternal X chromosome. For the carriers of anomalous vision the retina should have a mosaic of 4 cone types each with its own pigment. If these women use the information from their cones they must reject most trichromatic matches and only be satisfied with 4 variable, tetrachromatic matches. We find, however, that these carriers are always able to make a trichromatic match that satisfies them. However, we have discovered that for at least some of these women Grassmann's Laws do not hold. If we ask for a trichromatic match, and then add the same uniform background light to both sides of the matching field, the two sides of the field often become clearly distinguishable and the match has to be changed to a different setting (Fig. 3). The interpretation here is that in these carriers' retinas signals from normal cones and their anomalous replacements travel the same postreceptoral pathway and so become averaged, and that colored backgrounds selectively attenuate the signals from the normal or anomalous cones, whichever type is more sensitive to the background. The changes in the match follow the pattern to be expected on this hypothesis.

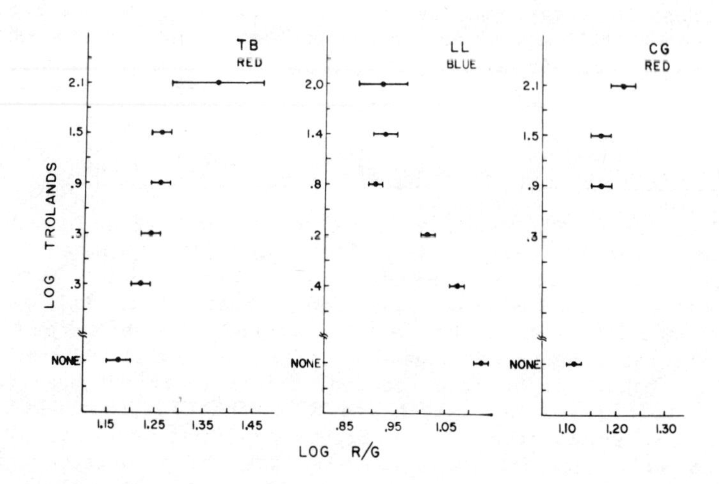

Fig. 3. Failure of Grassmann's Third Law (of additivity): the ratio of red (660 nm) to green (546 nm) primaries required in a mixture to match a 588 nm yellow (horizontal axis) is altered by red (670 nm) or blue (455 nm) backgrounds of various intensities in three women who were carriers of an abnormality of color vision.

The trichromatic vision of these women is therefore not the result of their having only 3 cone pigments but must be due to a trivariance of postreceptoral neural organization. Such a neural trivariance must be characteristic of normal visual systems too since neither the availability of an additional receptor type (the rods), nor variation in spectral sensitivity among cones of the same class, seems able to disrupt the trichromacy of steady-state normal visual matching. Although this neural trivariance must be recognized, a precise statement of what is implied by the concept involves difficult logical problems. To begin with, although it would be sufficient for neural trivariance if the receptors fell into only 3 classes on the basis of their central connections (as I have assumed here in the case of the anomalous carriers), that condition is not necessary and is doubtless not satisfied in the case of rod intrusion. It is more accurate to require that at some postreceptoral stage the various neural responses are dependent on only three variables which in turn are derived from a generally larger number of receptor signal variables. The simplest way for this situation to arise would be to have the three variables represented by univariant signals at some stage of processing. But even the most orderly physiological results

(Derrington, Lennie and Krauskopf, 1983) do not suggest an organization as tidy as this would require. Instead, postreceptoral variability from cell to cell is greater than at the receptor level. So long as this variability is present, it seems to me that we still do not have a satisfactory physiological basis for trichromacy; but we can say that under some conditions at least trichromacy rests on a neural (and not a receptoral) trivariance.

From Matching to Appearance

These reservations notwithstanding, it is agreed that for most observers under purely photopic conditions trichromatic matches are at least approximately determined by receptor constraints and will be practically unaffected by the vagaries of postreceptoral processing. But what about the appearance of the matched lights? To what extent do the receptors put constraints on color appearance? The short answer is, I think, not at all. If we think about this in physiological terms, we must realize that a visual neuron connected to 2 different cones with different pigments is logically free to exhibit any spectral sensitivity whatsoever (so long as the ratio of the quantum catches of the 2 cones is monotonic with wavelength). If 3 such neurons encode 3 dimensions of apparent color, we can say nothing at all a priori about how the firing rates or color sensations will depend on the cone quantum catches.

Even for a uniform field of light in the dark, color does not seem to be a simple fixed function of receptor excitations. One example: Rushton and Baker (1964) found that people vary substantially in their relative sensitivity to red and green and that this variation is correlated with their endowment of R and G cones as measured by retinal densitometry. Yet all such people see yellow at about the same point in the spectrum, with a far greater degree of consensus than Rushton and Baker's receptor measurements would lead one to expect (Rubin, 1961), suggesting that individual differences in neural organization induced genetically or (more likely) by experience compensate for the differences at the receptor level. Another example of this is the sensations of the unilaterally color blind. If color blindness in these cases is due simply to a change in the R or G visual pigments, it can be shown that the gamut of sensations they experience must be restricted to a vertical line in the (r, b) chromaticity diagram, running for instance from yellow to violet. This is not observed (MacLeod and Lennie, 1976), so an abnormality in neural organization must be modifying the appearance of colors, in addition to the influence of the pigment switch.

The Contribution of Receptors to Color Constancy

Color constancy is generally viewed as the prime example of a failure of color experience to correlate with receptor signals. I do not wish to argue that postreceptoral processes have no role in color constancy but I do think it is instructive to consider what is probably the simplest possible model for constancy, a model where constancy is essentially the work of the cones themselves.

William Rushton, during his last years, was very interested in the relationship between visual adaptation and color appearance. He had a model for adaptation in the rod adaptation pool and in the cone receptors, which actually implies color constancy, as well as the fading of stabilized retinal images which is what he was most interested in. In Rushton's adaptation model (never fully published, but see Rushton, 1972), a cone under an illumination of time average intensity $\bar{I}$ adopts a steady state sensitivity inversely proportional to $\bar{I}$ so that the output (intensity X sensitivity) is independent of $\bar{I}$. This, of course, immediately accounts for the fading of retinally stable images. When images are not stabilized each cone type will see temporal transients equal to the difference of excitation across the unstabilized contour. The signal produced will be proportional to the edge contrast as seen by that cone type. This is exactly what we need as the first step in implementing a Land type model (Land, this symposium) for color constancy.

Rushton (1972) got evidence that something like this does go on in cone vision in an experiment where he allowed a stabilized image of a bipartite field, with 2 half circles of different intensity I_1 & J_1 to fade into a uniform circle or disappear totally. Then he abruptly changed the intensities in the two halves of the field independently. In general, the field would now look non-uniform, but it could be kept uniform if the temporal intensity ratio (or temporal contrast, if you like) was made the same on the 2 sides of the field.

This experiment shows that adaptation to stable images introduces a reciprocal modulation of visual sensitivity, but it says nothing about the location where that change occurs. In fact, a series of many neural stages could be involved. I did some experiments to investigate this point by seeing whether the fading of stabilized images might occur independently in the rod and cone systems. If it does this would suggest that the processes concerned operate before the retinal ganglion cell level, where rod and cone signals come together. Like Rushton I used the Yarbus technique for image stabilization, with a suction cap contact lens stuck to the eye. Apparatus details aren't important except that the equipment provided two superimposed stabilized or unstabilized beams, one of short wavelength light for rods and the other of long wavelength for the cones. The essence of the results is most easily seen by considering one particular case: a bipartite field seen by rods, with the two halves different in intensity in a ratio of 4 to 1, was allowed to fade, and on it was superimposed a physically uniform flashed or unstabilized test field seen by cones. The key observation is simply that now with the transient test stimulus going through a different set of receptors than the stabilized background, the test stimulus that was physically uniform did appear subjectively completely uniform. The need for proportioning the intensities on the 2 halves to make the contrast the same, as found by Rushton, is no longer present. The processes that make the responses contrast-dependent occur separately in the rod and cone systems. I should note that this experiment leaves open the possibility that some processes relevant to fading occur after the ganglion cell, as physiological data tend to suggest; the implication is only that when stabilized images are involved such processes are more or less linear. Nor does the experiment locate the process as early as the rods and cones themselves necessarily: it could equally well be in the bipolar cells.

This receptor sensitivity modification is an outrageously simple model for image fading and color constancy and several objections could be raised against it.

First, adaptation may seem too quick to underly the fading of stabilized images. One could however defend the model by arguing that the adaptive state though rapidly modifiable crawls rather slowly toward its final asymptotic level, taking several seconds to achieve a reciprocity sufficiently perfect to prevent any visibility of stabilized image.

Land and Daw (1962) have shown that there can be approximate constancy in a flash. But again, most of the adaptation process is certainly very quick at ordinary light levels. And even if we need to invoke different processes to explain constancy in a flash, it is still possible that constancy in situations closer to the steady state is due largely to adaptation.

Adaptation is not all in the cones, as witness, for example, Dr. Pugh's contribution to this symposium. There are also color appearance phenomena, for instance the observation of supersaturated yellow after adaptation of the B cones which don't themselves respond to yellow - which call for explanation in terms of postreceptoral processes. This has to be acknowledged but I think we can still defend the Rushton model as a useful approximation. We don't yet know, very well, the relative importance of the post-receptoral factors under natural conditions. It's worth noting, I think that a point in favor of the "early fading" model is that when sensations are produced by direct electrical stimulation of the cortex, there is, according to Brindley (1973) no fading. Surprisingly, the cortex seems to lack the equipment for that job.

The resurrection of lightness and color from local contrast at nearby and remote contours in the field (Land, this symposium; Arend, Buhler and Lockhead, 1971) has to be imcorporated into a full constancy model, but the sensitivity regulation model is intended to describe only the first stage, the derivation of local signals suitable for later integration.

There are deviations from constancy: Helson (1938) found that in a room papered with neutral grays, whites and blacks the whites take on the illuminant color and the blacks take on the complementary color. This can very naturally be explained on the receptor model if we recognize that the simple intensity scaling model for adaptation is quantitatively inexact. There is independent evidence that the transient responses in the light adapted eye are more positively accelerated functions of stimulus intensity (if we represent them by power functions, they have a higher exponent) than in dark adaptation; and it has been shown by Nayatani, Takahama and Sobagaki (1983) that this will account well for the Helson deviations from constancy.

Finally, a more fundamental objection to this model, and to many others, is that it tries to explain constancy by making the signals produced under different illuminations the same. By doing that it does too much. We recognize illuminant quality as well as surface color, as pointed out most thoughtfully and convincingly by Heggelund (1974). We must assume that enough of a signal survives the adaptation process to provide cues to illuminant quality (and also, by the way to allow control of pupil size, which is known to be indifferent to image stabilization). Here again, it's not really clear whether the model is completely misguided or whether it can serve as a useful approximation.

In the light of all these objections the sensitivity regulation model has to be viewed as at best a rather crude and approximate account of the first stage in color constancy. It does have the merit of indicating a possibly important connection between constancy and the fading of stabilized images. It also is useful in illustrating the potential and limitations of what might be called "stupid" processes in color constancy - processes, in this case that produce signals simply related to local intensity differences and the time average of local intensity.

If we give up the idea that color constancy is achieved by the receptors, there is a rather wide range of alternative possibilities to explore. Going to the opposite extreme, the logically admissible arbitrariness in the relation between receptor signals and sensation is enough to allow perception to proceed like medical diagnosis: the signals from the cones, instead of being simply registered or subjected to simple more or less linear transformation might be invested with significance only as symptoms - symptoms of the presence of a certain kind of object surface. The information used could be quite varied including such things as awareness of the distribution of light sources, observation of the intensity profiles of shadows to exploit secondary reflection cues, remembered information from previous fixations, and so on. But a great problem with "smart" systems like this is that they require a lot of education to form the proper associations between receptoral symptoms and their objective causes - perhaps as many hours of education as a medical doctor gets. And we know from cases where people born with cataracts begin to see after cataract removal late in life (van Senden, 1960) that color naming can be very good immediately without refined visual experience to allow such learning. In view of this the potential of "stupid" systems— if not necessarily quite as stupid as the sensitivity modification system I've described--may still deserve to be explored.

References

Arend, L. E., Buehler, J. N. and Lockhead, G. R. (1971). Difference information in brightness perception. Perception and Psychophysics, 9, 367-370.

Boynton, R. M. (1980). Design for an Eye. In Neural Mechanisms and Behavior. (ed. D. McFadden). pp. 38-70. New York: Springer-Verlag.

Brindley, G. S. (1973). Sensory effects of electrical stimulation of the visual and paravisual cortex in man in Jung. (ed.). Visual Centers. In the Brain, Handbook of Sensory Physiology, 3, 583-594. Berlin: Springer-Verlag.

Dartnall, H. J. A., Bowmaker, J. K. and Mollon, J. D. (1983). Microspectrophotometry of Human Photoreceptors. In Colour Vision. (ed. J. D. Mollon and L. T. Sharpe). pp. 68-80. London: Academic Press.

Derrington, A., Lennie, P. and Krauskopf, J. (1983). Chromatic Response Properties of Parvocellular Neurons. In the Macaque LGN in Colour Vision. (eds. J. D. Mollon and L. T. Sharpe). pp. 245-251. London: Academic Press.

Eisner, A. and MacLeod, D. I. A. (1980). Blue-sensitive cones do not contribute to luminance, J. Opt. Soc. Am., 70, 121-122.

Granger, E. M. and Heurtley, J. C. (1973). Visual Chromaticity-modulation transfer function. J. Opt. Soc. Am., 63, 1173.

P. Heggelund (1974). Achromatic Color Vision - 1: Perceptive variables of achromatic colors. Vision Research, 14, 1071-1079.

Helson, H. (1938). Fundamental problems in color vision I. The principle governing changes in non-selective samples n chromatic illumination. J. Exp. Psychol., 23, 439-476.

Land, E. H. and Daw, N. W. (1962). Colors seen in a flash of light. Proc. Nat. Acad. Sci., 48, 1000-1008.

Luther, R. (1927). Aus dem Gebiet der Farbreizmetrik. Zeitschr. f. techn. Physik., 8, 540-558.

Luzzatto, L. and Gartler, S. (1983). Switching off blocks of genes. Nature, 301, 375-376.

MacLeod, D. I. A. and Boynton, R. M. (1979). Chromaticity diagram showing cone excitation by timuli of equal luminance. J. Opt. Soc. Am., 69, 1183-1186.

MacLeod, D. I. A. and Webster, M. A. (1983). Factors influencing the color matches of normal observers. In Colour Vision. (eds. J. D. Mollon and L. T. Sharpe). pp. 81-92. London: Academic Press.

Muller, G. E. (1924). Darstellung und Erklarung der verschiedenen Typen der Farbenblindheit. Vandenhoeck-Ruprecht, Gottingen.

Nagy, A. L., MacLeod, D. I. A., Heyneman, N. E. & Eisner, A. (1981). Four cone pigments in women heterozygous for color deficiency. J. Opt. Soc. Am., 71, 719-722.

Nayatani, Y., Takahama, K. & Sobagaki, H. (1981). Formulation of a nonlinear model of chromatic adaptation. Color Res. and Appl., 6, 161-171.

Rubin, M. L. (1961). Spectral hue loci of normal and anomalous trichromats. Am. J. Ophthalmol., 52, 166-172.

Rushton, W. A. H. (1971). Color vision: An approach through the cone pigments. Invest. Ophthal., 10, 311-322.

Rushton, W. A. H. (1972). Light and dark adaptation. Invest. Ophthal., 11, 503-517.

Rushton, W. A. H. & Baker, H. D. (1964). Red/green sensitivity in normal vision. Vis. Res., 4, 75-85.

von Senden, M. (1960). Space and Sight. London: Methuen.

Williams, D. R., MacLeod, D. I. A. & Hayhoe, M. M. (1981). Punctate sensitivity of the blue-sensitive mechanism. Vis. Res., 21, 1357-1375.

Stile's π_1 and π_2 Colour Mechanisms: Isolation of a Blue/Yellow Pathway in Normals and Dichromats

E.N. Pugh, Jr.,[1] J.E. Thornton,[2] L.J. Friedman[3] and M.H. Yim[4]

[1]Department of Psychology, University of Pennsylvania, Philadelphia, Pennsylvania 19104, USA
[2]Polaroid Research Corporation, Cambridge, Massachusetts, USA
[3]IBM Corporation, Philadelphia, Pennsylvania, USA
[4]University of Texas, Health Sciences Center, Houston, Texas, USA

INTRODUCTION

W. S. Stiles, in a series of classic studies (Stiles, 1939; 1949; 1953; 1959; 1978) developed the two-color increment threshold technique, and applied it to the analysis of the mechanisms of color vision. In Stiles's procedure an observer is adapted successively to each of a series of increasingly intense exposures of a large field of one wavelength (μ), and the threshold is measured for a relatively small and brief test of another wavelength (λ) presented in the center of the adapting field. The resulting data are plotted as an increment threshold curve or threshold-vs. intensity (t.v.i.) curve, giving $\log U_\lambda$, the logarithm of the threshold intensity as a function of $\log W_\mu$, the logarithm of the background intensity. Even if the test target is confined to the fovea, the full increment threshold curve usually is found to be composed of two or more distinct component regions or "branches", much as parafoveal dark adaptation curves normally comprise two distinct component regions.

In his 1939 masterpiece, "The directional sensitivity of the retina and the spectral sensitivities of the rods and cones", Stiles pointed out that the component branches of the two-color t.v.i. curves obey two displacement laws. The first law, the Test Displacement Law, states that changing the wavelength of the test probe causes the components observed upon a given adapting field to shift vertically (to varying extents), but causes no lateral shifting. The second law, the Field Displacement Law, states that changing the wavelength of the adapting field causes any component branches observed with a given test to move horizontally, but causes no vertical shifts. In other words, the component branches' shapes in log/log coordinates are invariant over changes in wavelength of test or adapting fields, and the effects of changing test and field are independent. The test spectral sensitivity of a component branch is a graph of the reciprocal absolute threshold of the branch plotted as a function of the test wavelength. The field spectral sensitivity is a graph of the field intensity at each

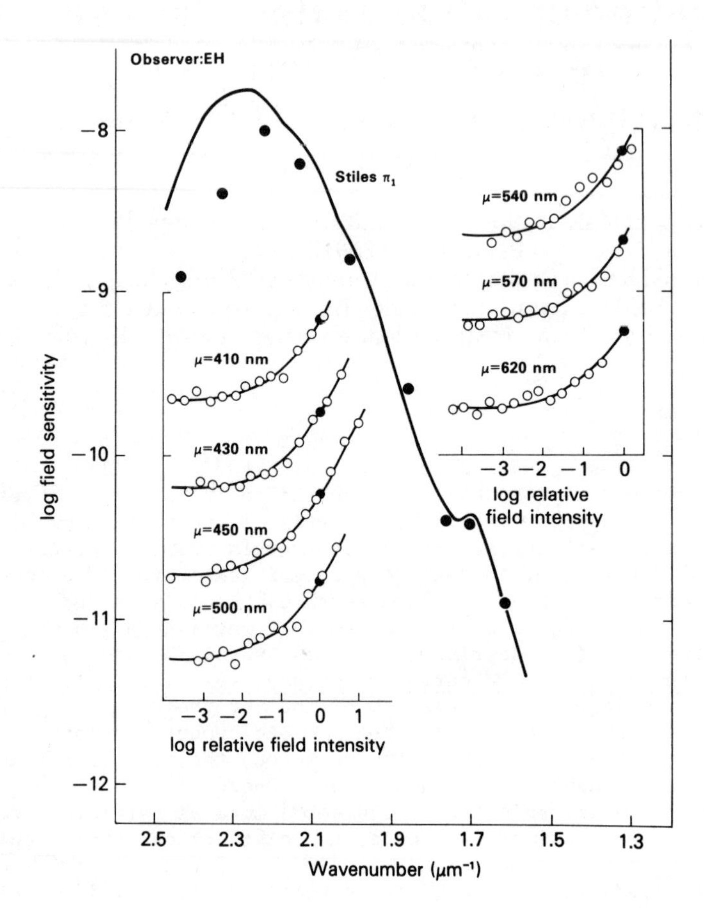

Figure 1.

Insets, open symbols. Foveal increment threshold curves for observer EH for 450 nm test target on 10 deg backgrounds whose wavelength is given by the parameter labelling the curves. All thresholds of data reported in this paper were measured with a temporal two-alternative forced choice staircase procedure. The test target was the "low frequency" test of Thornton & Pugh (1983): in space it is approximately a radially symmetric Gaussian with 3 deg width at half peak; in time it is a raised cosine of trough-to-trough duration 500 msec. (In subsequent figures the intensity units for thresholds measured with the low-frequency test are quanta.sec^{-1}.deg^{-2} at the peak in space in time.) The curves are placed on the logarithmic abscissa so that the point of 1 log unit threshold elevation (●) is at 0. A 590 nm auxiliary field of $10^{7.6}$ quanta.sec^{-1}.deg^{-2} was always present for the t.v.i. curves.

Filled symbols. Field sensitivity curve derived from the t.v.i. curves shown in the insets. (T.v.i. data for field wavelengths 470 nm and 590 nm are given in Figure 2). Solid curve is Stiles's average π_1 field sensitivity slid vertically for coincidence with filled symbols in the long wavelengths. Ordinate unit is -log (quanta.sec^{-1}.deg^{-2}).

---

wavelength required to raise the threshold of a component branch 1 log unit.

Figure 1 (insets, open symbols) shows a family of foveal increment threshold curves obtained this past year from an observer in our laboratory, illustrating the Field Displacement Law, and derivation of a field sensitivity curve. The test flash wavelength was always 450 nm. The standard Stiles increment threshold template (Wyszecki & Stiles, 1982, p. 532) was fit to each curve; the log reciprocal of the field intensities causing 1 log unit threshold elevation is plotted as the filled symbols.

In 1939 Stiles hoped that the two-color threshold technique could measure the spectral sensitivities of the three classes of foveal cones. This hope, however, gave way as more data were collected in the 40's and 50's. By 1953 Stiles found that full analysis of his foveal t.v.i. data required the postulation of five and possible seven distinct component branches. In particular, he found that in most observers there were three component branches whose test and field sensitivities peaked in the short wavelength region of the spectrum. Because the very existence of 5-7 curves with distinct spectral sensitivities precluded the possibility that each branch was governed by a different cone class, Stiles adopted a more agnostic analysis and called the component branches "π-mechanisms" (a name he told me that was not chosen for any special symbolic significance). In Figure 1 the average field sensitivity of Stiles's π_1 component branch is plotted along with the field sensitivity of our observer EH. The shapes of the two curves are in fair agreement, especially over the mid- and long-wave part of the spectrum, although in fact EH is about 0.5 log units more field-sensitive at peak than Stiles's average observer. Despite this absolute discrepancy and the difference in test parameters from those of Stiles (see caption), we feel confident that the branch isolated in EH is a manifestation of the same physiological substrate as Stiles's π_1. (π_1 field sensitivity data of two other observers studied in our lab may be found in Pugh, 1976.)

STILES'S π_1 and π_3 MECHANISMS

In this paper we review our efforts over the past decade to characterize the π_1 and π_3 mechanisms of Stiles. In our approach to explaining these component curves we have clung to two premises. The first is that normal foveal color vision is initiated by quantal absorption in three spectrally distinct cone classes, whose properties are such as to account for Grassmann's Laws of Metameric Color Matching (Krantz, 1975; Wyszecki & Stiles, 1982). The second premise is that in isolating a π-mechanism in the two-color increment threshold paradigm one is in fact isolating a stable physiological substrate, a well-defined *pathway*, originating in one or more of the cone classes and transferring information to the higher visual centers. This approach demanded that we devise experiments that test hypotheses about the relationship of each π-mechanism to the three cone classes, and that an explanation of each π-mechanism be developed explicitly in terms of signals from the three cones.

A useful way of conceptualizing Stiles's operational mechanisms is to think of the test flash as generating a linear perturbation signal which is attenuated by a gain factor set by the adapting field. Letting A represent an adapting field of arbitrary spectral distribution (an infinite dim. vector), $U_\lambda(A)$ the threshold for a test flash λ superimposed on A, and gi the *gain function* of the ith component branch, then

$$U_\lambda(A)/U_\lambda(0) = 1/g_i(A) \qquad (1)$$

where $U_\lambda(0)$ is the threshold in the absence of a field. By virtue of Trichromatic principles (Stiles, 1967; Krantz, 1975) gi must be a function only of the three cone quantum catch rates: $g_i(A) = G_i[\alpha(A), \beta(A), \gamma(A)]$, where $\alpha(A)$, $\beta(A)$, $\lambda(A)$ represent the quantum absorption rates of the short-wave, mid-wave, and long-wave cones respectively from the field A. For a mechanism to have a well defined field sensitivity, Gi must be expressible as a function only of this sensitivity multiplied by the field intensity. Thus, if the field is monochromatic of wavelength μ and intensity W_μ, and if the t.v.i. curves follow the Stiles standard shape, $G_i[\alpha(W_\mu), \beta(W_\mu), \lambda(W_\mu)] = \zeta(\Pi i_\mu W_\mu)$, so that Πi_μ must express the intensity-invariant contribution of the cone signal(s) controlling the adaptation state of the branch.

The simplest hypothesis about the relationship of a component t.v.i. branch to the cones is that all threshold elevation of the branch is effected by signals of one cone class. Under this hypothesis one may deduce another Stilesian Law that the branch must obey, the *Law of Field Additivity*. This law requires that the effect of any bichromatic field in elevating the threshold of a component branch is equal to the effect of the sum of the field-sensitivity-weighted component fields. For a mechanism having the Stiles gain function ζ, Field Additivity can be succinctly expressed as

$$g_i(W_{\mu_1} \oplus W_{\mu_2}) = \zeta(\Pi i_{\mu_1} W_{\mu_1} + \Pi i_{\mu_2} W_{\mu_2}) \qquad (2)$$

where $W_{\mu_1} \oplus W_{\mu_2}$ represents the physical superposition of the fields. Thus, the threshold on any bichromatic field for a component branch obeying Field Additivity is completely determined (and may be readily calculated) if one knows the effectiveness of the individual

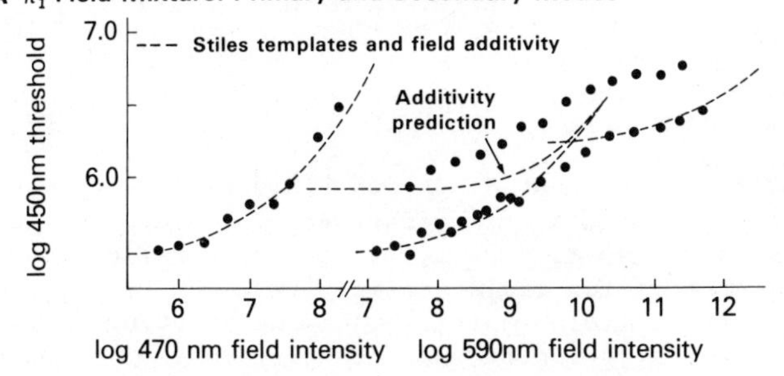

B π_1 Steady-state Adaptation (Pugh, 1976)

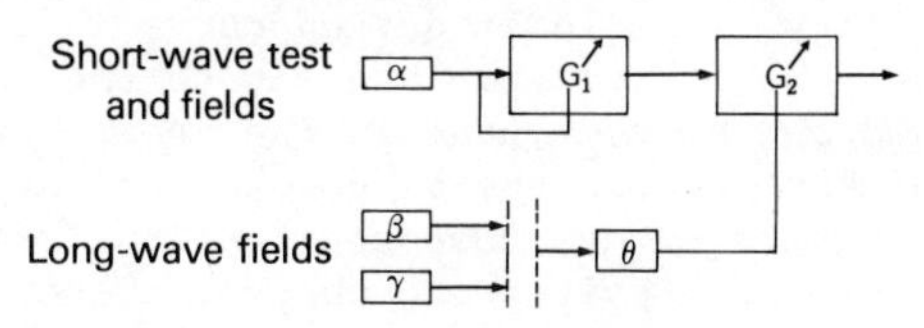

C π_1 Dynamics (Augenstein and Pugh, 1977)

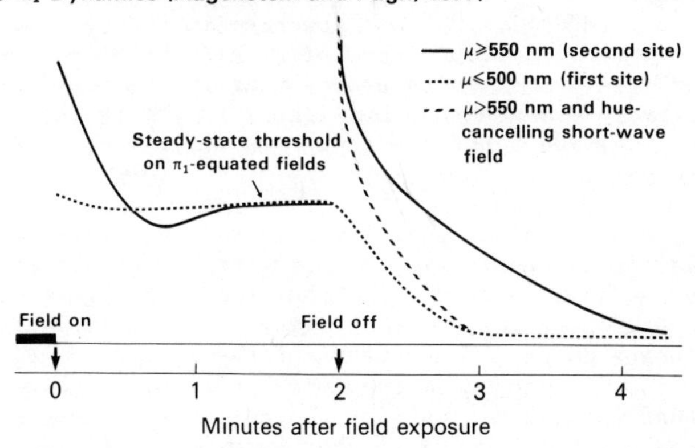

Figure 2.

A. π_1 field-mixture experiment. Left panel: t.v.i. curve on 470 nm field alone. Right panel: lower set of data, t.v.i. curve on 590 field alone; upper set of data, thresholds on mixtures of same 590 nm fields with 470 nm field of $10^{7.6}$ quanta.sec.$^{-1}$.deg^{-2}. Lower dotted curves, Stiles ζ-template. Upper dotted curve in right panel, prediction of Field Additivity. Field intensities in quanta.sec^{-1} deg^{-2}. "Low frequency" test (see caption, Figure 1). Observer EH.

B. Two-site gain theory of π_1 pathway proposed by Pugh (1976).

C. Dashed curve time course of π_1 adaptation for $\mu < 500$ nm. Solid curve: time course of π_1 adaptation for $\mu > 550$ nm. Note that both curves show the same steady-state threshold elevation. Intermittently dashed curve: recovery time course at extinction of long-wave field when bright bluish "cancelling" field is admixed. Curves summarize data in Augenstein & Pugh (1977).

---

monochromatic fields.

In 1975 we found that π_1 obeyed Field Additivity when the two monochromatic components of a bichromatic mixture were both of wavelengths less than c. 500 nm, (in the spectral neighborhood of the primary mode of the field sensitivity), but that π_1 systematically failed to be field-additive when one adapting field component was taken from the region of the secondary mode ($\mu \geqq 550$ nm) and the other from the region of the primary mode ($\mu \leqq 500$ nm) (Pugh, 1976). Figure 2A illustrates, with data from EH, this particular failure of π_1 to obey Field Additivity. The lower dotted curve in each half of the figure represents the Stiles template fit to the individual monochromatic 470 nm and 590 nm components. The upper set of points in the right-hand panel are the thresholds upon bichromatic fields composed of a fixed-intensity 470 nm component and the same series of 590 nm fields: if Field Additivity were obeyed, these points would follow the predicted upper dotted line. The thresholds on the bichromatic fields rise much sooner than predicted; the mixture fields are "superadditive." Upon the premise that a π-mechanism is the manifestation of a stable physiological substrate, the failure of π_1 to obey Field Additivity proved that the adaptation events in this substrate are governed by signals from more than one class of cones (Pugh, 1976). But how could one explain π_1 in terms of the three cones?

Since π_1 has a unique test sensitivity that peaks at 430-450 nm and falls off in sensitivity by more than 1 log unit toward 500 nm, we adopted the working premise that the test signal detected by an observer "operating on π_1" originates exclusively in the short-wavelength cones. The pattern of the failure of Field Additivity itself suggested a specific hypothesis, shown in Figure 2B, viz., that the adaptation events occur in a pathway with two successive gain sites. When two short-wavelength fields are mixed their effects summate according to the Additivity Law because both lights only affect the first site, whose gain is governed by absorptions only in the short-wave-sensitive cones; when a short- and a long-wave field are combined the adaptational effects (gains) are multiplied, since the two fields affect distinct serial gain sites.

If two physiologically distinct sites are involved in setting the gain of the detection pathway governing threshold under π_1-

isolation conditions, then there is no reason that the two sites should have identical adaptational dynamics: in other words, the time course of adaptation to short- and long-wavelength fields equated for their steady-state π_1 threshold elevations might well be expected to be different. This is exactly what we discovered (Augenstein & Pugh, 1977). Figure 2C schematically summarizes the result. The very fields ($\mu \geqq 550$ nm) which steady-state Field Additivity failures dictated to be adapting the pathway at the second site gave rise to strikingly different adaptation dynamics than the fields ($\mu \leqq 500$ nm) the steady-state analysis said were adapting the first site. The large transient elevation in π_1 threshold occurring at the extinction of long-wave adapting fields was first observed by Stiles (1949), and has been studied by John Mollon and colleagues (see, for a summary, Mollon & Polden, 1977a), who named the phenomenon "transient tritanopia" and described its properties in rich detail.

In 1976 (Augenstein & Pugh, 1977) we discovered a striking and revealing feature of transient tritanopia: admixing a fairly intense bluish component to the long-wave fields that would ordinarily cause transient tritanopia at their extinction, eliminated most of the phenomenon -- despite the fact that the short-wave addend necessarily produced a large "first-site" adaptation. This observation, illustrated schematically in Figure 2C, led to the conclusion that the second site must be "cone-antagonistic" -- i.e., its gain was set by its response to an input that combined oppositely signed signals from the short- and from the mid- and long-wavelength cones. A critical steady-state field-mixture experiment demonstrating directly the cone-antagonistic nature of the second site was reported in 1977 by Polden & Mollon; Figure 3 shows a replication of their experiment (see also Pugh & Larimer, 1980; Thornton & Pugh, 1983; Friedman, Yim & Pugh, 1984 for further replications). The choice of a c. 470 nm adapting field allowed Polden & Mollon to follow the π_1 threshold over a wide range of field intensities. They observed directly, as is shown in Figure 3, that admixture to intense 470 nm fields of the yellowish fields responsible for the secondary mode of the π_1 field sensitivity causes the threshold to drop! It is thus patent that the hypothesized second site must set its steady-state gain on the basis of a cone-antagonistic input: and that the second site has the property that it is most sensitive under approximately chromatically neutral adaptation conditions.

In 1979 an explicit mathematical formulation of the two-site, cone antagonistic model of the π_1 pathway was presented (Pugh & Mollon, 1979). For an arbitrary adapting field A this model has gain given by

$$g(A) = \zeta_1[K_0\alpha(A)] \cdot \zeta_2[|K_1\alpha(A) - K_2\beta(A) - K_3\gamma(A)|] \qquad (3)$$

Here $\zeta i[.]$ is the standard Stiles gain function (of site i), $\alpha(A)$, $\beta(A)$ and $\gamma(A)$ are the quantum absorption rates of the three cones, and the Ki's parameters. (In Pugh & Mollon (1979) a slightly more

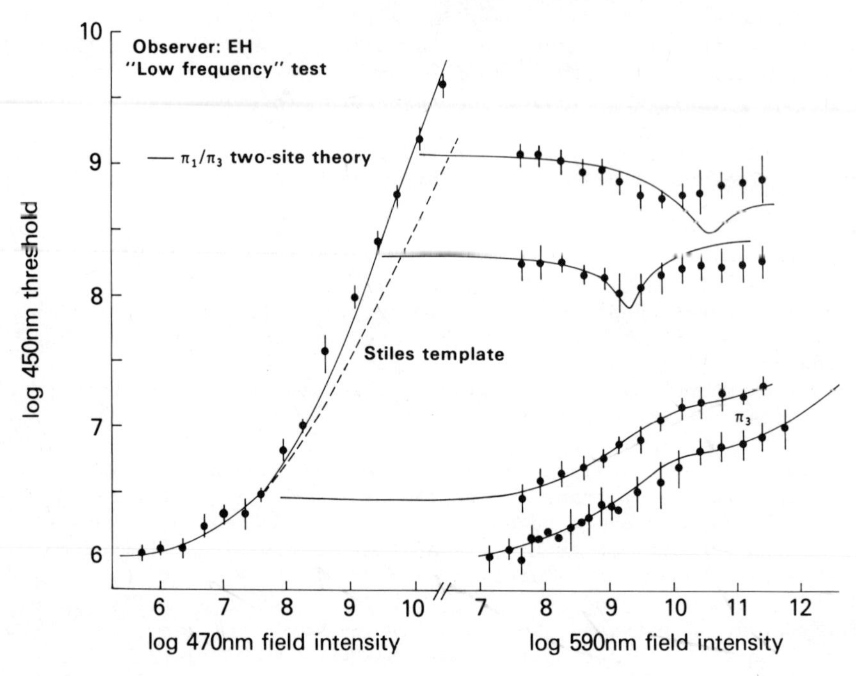

Figure 3.

π_1 field mixture experiment for observer EH. Left panel: t.v.i. on 470 nm field alone. Right panel: thresholds on mixtures of 590 nm fields admixed to four different 470 nm fields, of intensities 0, $10^{7.6}$, $10^{9.4}$, $10^{9.95}$ quanta.sec^{-2}.deg^{-2}. The test was the "low frequency" test (see Figure 1, caption). Each point is the mean of at least two replications across days; error bars represent +/- 2 s.e.m.

The solid lines were computed with equations (1) and (3) of the text. The parameters were $K_0 = 10^{-8.0}$, $K_1 = 10^{-9.7}$, $K_2 = K_3 = 10^{-10.15}$ reciprocal quanta.sec^{-1}.deg^{-2}. The "half-bleaching" constant for long- and mid-wave cone pigments used was $10^{9.6}$ quanta.sec^{-1}.deg^{-2}.

---

general, nonlinear formula for the input to the second-site gain function was used.) This model was shown to recover the major qualitative effects known about π_1 with reasonable parameters, as demonstrated in Figure 3, where all the solid lines drawn through the data here are theoretical curves computed with the equations (1) and (3). Upward deviation from the Weber line on the 470 nm field occurs in the theory because blue adapting fields cause simultaneous adaptation at both first and second sites -- first site adaptation alone gives limiting Weber behavior (dotted line in left panel; see Figure 4). π_1 superadditivity (see also Figure 2A) occurs when long-wave fields are admixed to low intensity

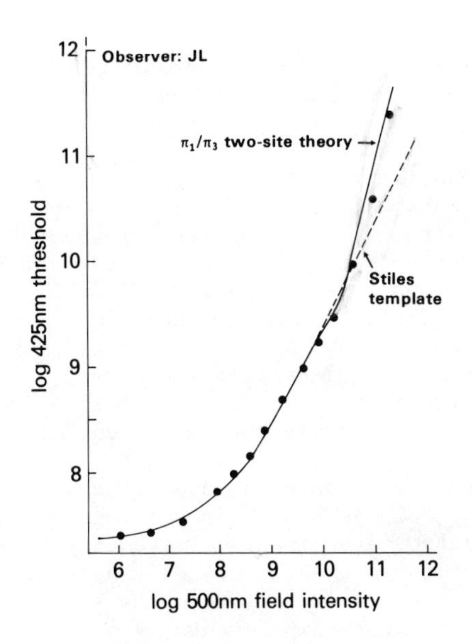

Figure 4.

π_1 t.v.i. curve for observer J.L., standard Stiles conditions -- 230 msec, 1 deg foveal test. Data from Pugh & Larimer (1980, Figure 8). Dashed line, Stiles template. Solid line computed with 2-site gain theory, equation (1) of Pugh & Mollon (1979): $K_0 = 10^{-8.0}$, $K_1 = 10^{-10.0}$, $K_2 = K_3 = 10^{-10.6}$ reciprocal quanta.sec^{-1}.deg^{-2}; n = 0.69 (nonlinear power coefficient); $W_0 = 10^{9.7}$ quanta .sec^{-1}.deg^{-2} ("half-bleaching" constant).

---

short-wave fields because at low intensities the short-wave fields adapt only the first site, the long-wave only the second, and the short-wave fields are not sufficiently intense to affect the second site polarization. When the same long-wave fields are mixed to intense short-wave adapting fields, however, cancellative sub-additivity takes place at the second site: the thresholds fall down approximately to the Weber line. The theory gave gratis an explanation of the "limited conditioning effect" or π_3 plateau -- because as the intense steady-state long wave fields come into the bleaching range of the long- and mid-wave cones the signals from the long and mid-wave cones necessarily must approach asymptote (Augenstein & Pugh, 1977). (Operationally, of course, any process that would cause the β- and γ-cone signal to the second-site would give the same result. Empirically, for some observers the requisite saturation seems to occur up to a log unit too soon to be due to bleaching.) The π_3 branch observed on the long-wave fields then can be explained as the adaptation by the long wave fields of first site, which because its spectral sensitivity is that of the short-wave cones, requires extremely intense light at such wavelengths (Pugh & Mollon, 1979).

Figure 4, a π_1 increment threshold curve on a 500 nm field, serves to illustrate further some interesting behavior of the two-site theory. Note first how the π_1 t.v.i. data follow the Stiles template (dashed line) on the 500 nm field up to about 10^{10} quanta $.sec^{-1}.deg^{-2}$: in equation (2) this occurs because $K_1\alpha(W_{500}) - K_2\beta(W_{500}) - K_3\gamma(W_{500}) = 0$, the second-site remains unpolarized, and only the first site adapts. However, at about $10^{10.5}$ quanta $.sec^{-1}.deg^{-2}$ the data suddenly leap above the Weber line: theoretically this occurs because the β- and γ-cone signals to the second site saturate (e.g., due to bleaching), and the $K_1\alpha$ signal to the second site overtakes $K_2\beta + K_3\gamma$, polarizing the second site and reducing its gain. The theory (solid curve) actually predicted the shape and position of this striking deviation from Weber behavior more than a year before the experiment was performed.

We have performed various additional experiments to further characterize what we felt justified in calling in 1980 the "π_1/π_3 pathway." In one set of experiments (Friedman, 1982; Friedman, Yim & Pugh, 1984) we found that in the protanope and in the deuteranope, π_1 behaves essentially the same way as in normal trichromats in field mixture experiments -- viz., apparently has a second site gain that is determined by a cone-antagonistic signal. Figure 5 shows the results we obtained from one deuteranope, and also shows that essentially the same behavior in the field-mixture experiment is observed whether the test target is the standard Stiles 200 msec, 1 deg target, or our low-frequency test.

Another interesting feature of the pathway is that its temporal integration does not depend on its state of adaptation (Friedman, Yim & Pugh, 1984). Figure 6 illustrates this behavior, and shows how it can be used to diagnose intrusions of other mechanisms. The upper panel gives the field-mixture data of a normal trichromat; the lower panel gives seven time x intensity tradeoff functions obtained at six different states of adaptation of the pathway, as indicated by the letter A-F in the upper panel. A shows the I x T tradeoff in the presence of a weak 470 nm field; B, C, D for more intense 470 nm fields to which various 590 nm fields have been admixed. The critical duration -- defined as the intersection of the slope -1 segment with the slope 0 segment -- is about the same for the conditions A-D. At E, however, the critical duration for the 425 nm flash appears anomalously short, suggesting the intrusion of another mechanism for the detection of the short flashes. At F, a two-branched I x T tradeoff function is observed for the 425 nm test. The slower component matches that attributable to the π_1/π_3 pathway under other adaptation states, and the threshold for a 526 nm flash under the same adapting condition (F') shows the same behavior as the faster component in the 425 nm curve. Thus,

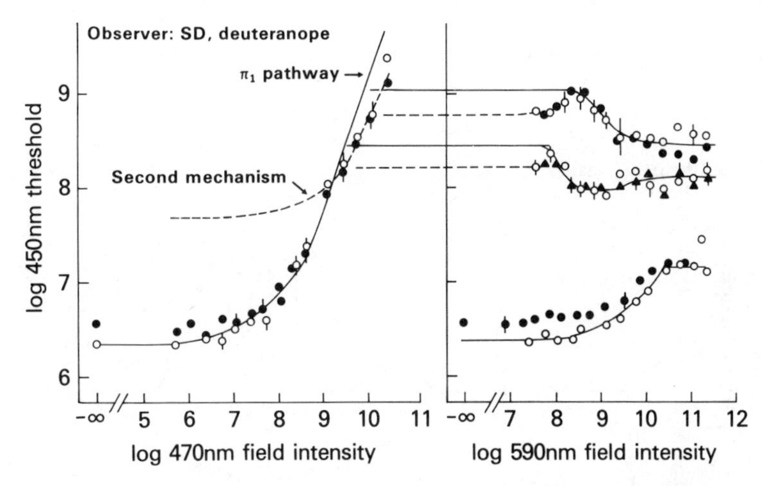

Figure 5.

Foveal field-mixture experiment in deuteranope showing cancellative subadditivity of fields. Open symbols: "low frequency" test (see caption, Figure 1). Filled symbols: 200 msec, 1 deg test. Solid curves represent the π_1/π_3 pathway thresholds (not computed with theory). On 470 nm fields of intensity greater than 10^9 quanta.sec^{-1}.deg^{-2} another mechanism (exhibiting Weber behavior) and field additivity with the 590 nm fields appears to intrude. Critical duration experiments confirm this analysis (Friedman, 1983; Friedman, Yim & Pugh, 1984).

---

we conclude that a mid-wave sensitive mechanism is intruding in detection of brief flashes at E and F. Essentially the same analysis applies to the deuteranope data of Figure 5 (Friedman, Yim & Pugh, 1984), and the curves in Figure 5 have been drawn to indicate the presence of the mid-wave mechanism.

CONE-ANTAGONISM OF π_1/π_3 PATHWAY COMPARED WITH CONE ANTAGONISM OF OPPONENT COLORS' BLUE/YELLOW CHANNEL

Up until this point we have avoided the use of color names in characterizing the π_1/π_3 pathway. And this is entirely appropriate, for all the data in Figures 1-6 were collected with threshold procedures. Nonetheless, the cone-antagonistic site of the pathway bears more than a passing resemblance to Opponent Colors Theory's cone-antagonistic model of the Blue/Yellow channel (Hurvich & Jameson, 1957), and from the earliest identification of the cone-antagonism of the second-site we were interested in the relationship between the threshold cone antagonism we had found and that postulated by Hurvich & Jameson's Opponent Colors Theory to account for the properties of blue/yellow hue equilibrium data.

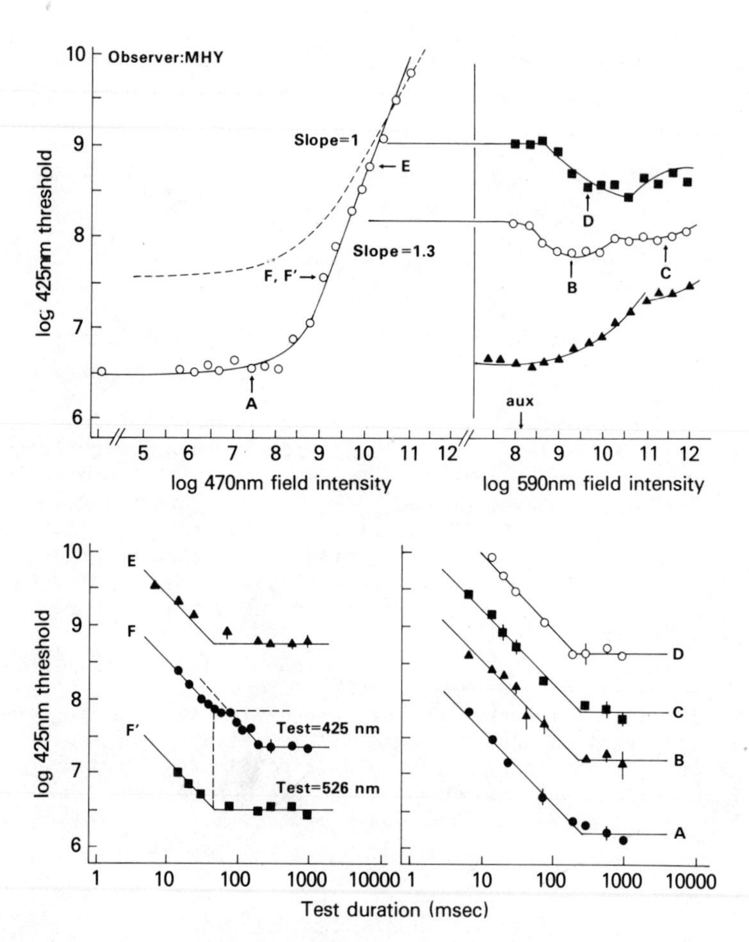

Figure 6.

Upper panel: field-mixture experiment for observer PY. Test: 200 msec, 1 deg, foveal.

Lower panel: I x T tradeoff experiments. A-F label adaptatio conditions indicated by the same letter in the upper panel. Curve A and B are shifted downward by 0.2 and 0.5 log units respectively F' represents the same adaptation condition as F, with test wave-length changed to 526 nm.

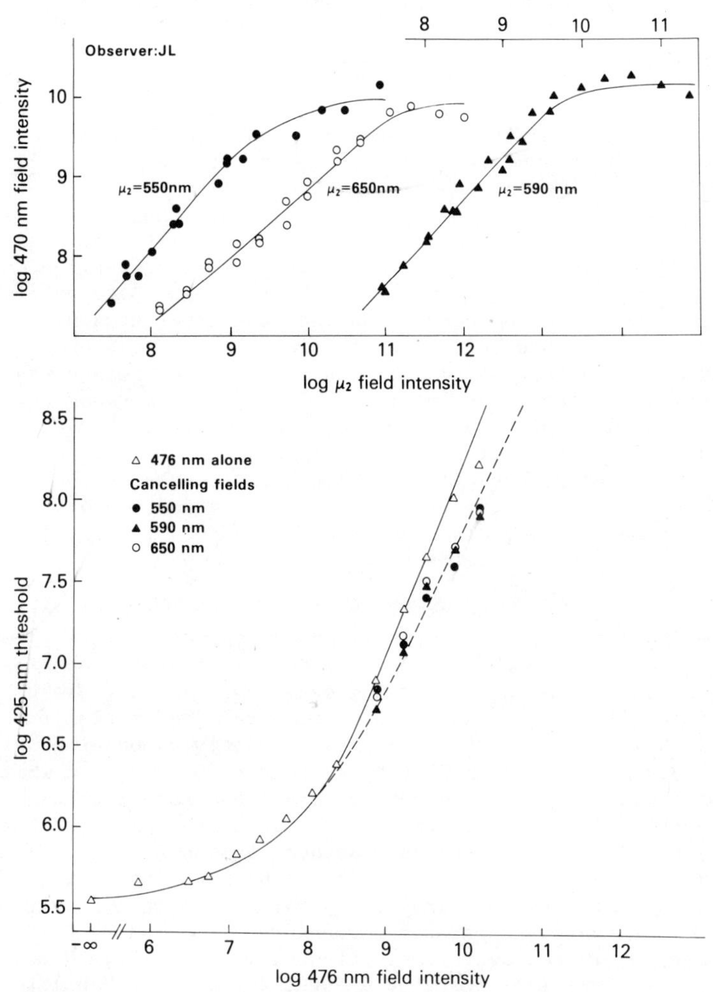

Figure 7.

Upper panel. Yellow/blue hue equilibria for 10 deg, steady-state fields for observer JL. Lower abscissa gives the intensity of 550 nm or 650 nm component; upper abscissa of 590 nm component. Ordinate gives intensity of yellow cancelling 476 nm field component. Procedure described in Pugh & Larimer (1980).

Lower panel. Filled symbols give t.v.i. curve for 425 nm, 200 msec, 1 deg foveal test on 476 nm fields. Open symbols give thresholds upon same fields brought to blue/yellow equilibrium by admixture of 550 nm (triangles), 590 nm (circles) or 650 nm (squares) yellowish fields derive from data of upper panel. Yellowish field components cancelling the same blue component cause equal cancellative subadditivity effect, i.e., same increase in second-site gain in the pathway.

Figure 7 shows data from our initial effort to compare quantitatively the cone-antagonism of the second-site gain of π_1/π_3 pathway and that postulated by Opponent Colors Theory (Pugh & Larimer, 1980). As shown in Figure 7A, we found, for each of series of increasingly intense 476 nm fields, the 650 nm, 590 nm or 540 nm component which brought the mixture to blue/yellow hue equilibrium. We then measured the thresholds upon the equilibrium mixtures, as shown in Figure 7B. For two observers we found that second site adaptation for all these combinations was equivalent and minimized. In other words, the second site gain was maximized when the mixture field appeared neither blue nor yellow. Although these observations were consistent with the conclusion that the cone-antagonism of π_1 and the cone-antagonism of Opponent Colors Theory were one and the same, the outcome was not totally satisfying. One reason was that Mollon tried the same experiment and did not get the same result -- although his threshold experiments were not done with 2AFC procedures, and his hue equilibria were not measured in quite the same fashion. Another reason was that threshold troughs in the field-mixture experiments (e.g., Figure 3) are broad and roughly maximal gain of the second site also occurs over a range of slightly bluish and yellowish adapting fields.

A breakthrough in the problem of comparing the cone antagonism of the π_1 pathway with that postulated by Opponent Colors Theory for its Blue/Yellow pathway came in a series of experiments on long-wave/mid-wave cone antagonism using our "low frequency" test, whose energy is confined to relatively low temporal and spatial frequencies. Figure 8A shows a test spectral sensitivity obtained with this stimulus for observer EH adapted to a 6000 td "xenon white" adapting field. The data show three prominent modes: the long- and mid-wave modes are too narrow and peak at the wrong wavelengths to be identified with the cones. Rather, as shown previously by Sperling & Harwerth (1971), these data require a cone-antagonistic ("opponent") interaction. The theory fit to the full spectrum in fact has four cone-antagonistic components, labelled B, G, Y and R: each component is a linear combination of the small-field color matching functions (see Thornton & Pugh, 1983a). Probability summation is used to combine signals in the regions of overlap. Although the theoretical fit is not so good as that for the data of other observers under the same or similar conditions (Thornton & Pugh, 1983a,b), the labels applied to the mechanisms have a significance not derived from the spectral fit, as we now show. Figure 8B shows the result obtained when thresholds for constant-ratio mixtures of 620 nm and 530 nm flashes are measured under the same conditions: here the cone-antagonistic interaction of signals from the β- and γ-cones can be observed directly. The two parallel straight lines with positive slope represent linear cone-antagonistic R and G mechanisms; the negatively sloped line represents another linear mechanism, and the curvilinear line represent the combination, by probability summation, of the three mechanisms. In Figure 8C the two straight lines of 8B are plotted in logarithmic coordinates and absolute units. The four open symbols represent

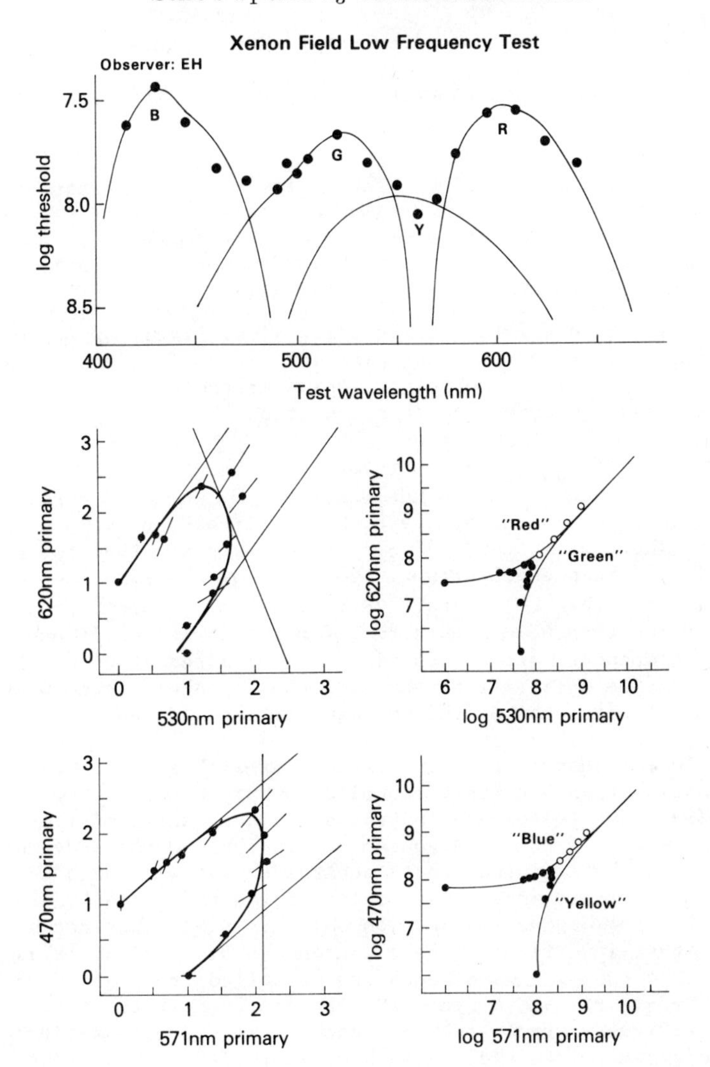

Figure 8.

A. "Low frequency" test spectral sensitivity for observer EH adapted to 6000 td "xenon white" field. Curves through data are described in the text.

B. Test-mixture experiment for EH. Thresholds were measured for various constant-ratio mixtures of 530 nm and 620 nm "primaries" using a temporal two-alternative forced-choice staircase procedure: each constant-ratio is represented by a ray emerging from the origin, and thus the error bars (2 s.e.m. across days) lie along these rays. In the theoretical fit the positively sloped lines are constrained to be parallel.

C. Test-mixture data (filled symbols) of Figure 8B replotted in logarithmic coordinates. The parallel lines of 8B are also replotted in the same coordinates. Open symbols are supra-threshold red/green hue equilibria, determined with double random-staircase procedure.

D. Test-mixture experiment for EH for various constant-ratio mixtures of 470 nm and 560 nm "primaries". The 470 nm flash alone is detected by the π_1/π_3 pathway; the 560 nm primary thus specifically inhibits detection through this pathway. Positively sloped lines constrained in fit to be parallel.

E. Test-mixture data (filled symbols) of Figure 8D replotted in logarithmic coordinates. The parallel lines of 8D are likewise replotted. Open symbols represent supra-threshold blue/yellow hue equilibria measured under the same conditions.

---

supra-threshold hue red/green hue equilibrium judgments obtained under the same conditions with a double staircase on the 620 nm component (Thornton & Pugh, 1984). It thus appears that the supra-threshold red/green equilibrium is governed by the same components that govern the threshold data. "Red" and "Green" are appropriate labels for the threshold mechanisms, for the dominance of one or the other component corresponds with the color sensation of the same label. The spectral signature of the threshold cone-antagonism is consistent with that of the hue equilibrium judgments.

The low frequency test has provided us with another way -- more sensitive than the field cancellation procedure of Pugh & Larimer (1980) -- to compare the threshold cone-antagonism of the π_1 pathway with the cone-antagonism of blue/yellow hue judgments. Moreover, the low frequency test spectral sensitivity data (Figure 8A) told us where we would have to look for a threshold "yellow" signal, and provided an explanation why this signal has not been found by other investigators (e.g., Kranda & King-Smith, 1979). We know that the two long-wave humps can be called "red" and "green". Thus, if we are to find a "yellow" signal at threshold we must look in the relatively narrow spectral range where the mid- and long-wave cone signals into the R/G pathway annihilate one another. Figure 8D shows the data obtained when a test wavelength from between the R and G humps is combined with one from the short-wave hump, and thresholds measured for various constant-ratio mixtures measured: a clear and robust cone-antagonism is found. Since the detection of the short-wave component of the threshold mixture can be unambiguously assigned to the π_1 pathway on the basis of the standard criteria of test and field sensitivity, it is clear that the perturbation signal initiated by the 570 nm flash specifically opposes the short-wave cone signal into the π_1 pathway initiated by the 470 nm component. When long-wave flashes not taken from the region between the R and G peaks of the spectral sensitivity are used, the antagonism is not observed (Thornton & Pugh, 1982, 1984). Figure 8E shows the threshold data and parallel lines of

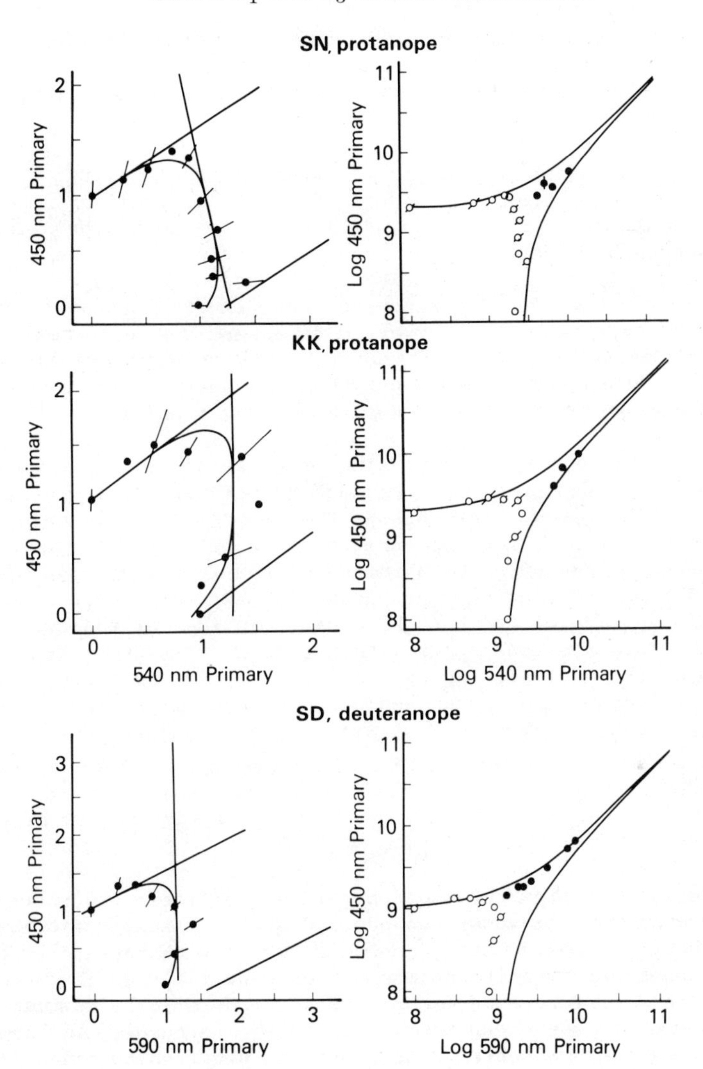

Figure 9.

Left panel: threshold test-mixture experiments for 2 protanopes adapted to approximately 20,000 td "xenon white" adapting field and 1 deuteranope adapted to color neutral monochromatic field of wavelength 503.4 nm and intensity $10^{10.6}$ quanta.sec^{-1}.deg^{-2}. For protanopes the mixtures were of 450 nm and 540 nm; for the deuteranope, 450 nm and 590 nm. Low-frequency test used in all cases. Theoretical curves fit to the threshold data were constrained to be consistent with the supra-threshold blue/yellow hue equilibrium data.

Right panel: threshold data of left panel replotted in logarithmic coordinates, along with supra-threshold hue-equilibrium judgments. Because of the intensity of the adapting fields, relatively little intensity range was available supra-threshold due to apparatus limitations.

8D replotted in logarithmic coordinates and absolute units. The open symbols show supra-threshold blue/yellow hue equilibrium judgments by EH under the same conditions. To a good approximation the supra-threshold data are governed by a pathway with the same cone-antagonistic signature as that which governs the threshold inhibitory effects. We have repeated these experiments on several other normal trichromatic observers with essentially the same results, although in some cases the threshold cone antagonism can be followed out as far as 300%, and thus its spectral signature in the mixture experiment determined with greater precision.

Figure 9 shows similar data similar to those of Figure 8D and 8E obtained from two protanopes and one deuteranope (Friedman, Thornton & Pugh, 1984). For reasons that we do not fully understand we have not been able to achieve the same degree of isolation in dichromats as in normal trichromats of the long-wave side of the cone-antagonistic pathway at threshold, but the signature of the cone antagonism of the threshold data and that of the hue equilibrium data are consistent within the sensitivity of the estimates of the best fitting parallel lines. The relative strength of the B/Y opponent and the non-opponent mechanism in dichromats however, apparently differs materially from the relative strength in normals.

SUMMARY

Stiles's two-color threshold provides a reliable procedure for isolating what we believe are unique and stable visual "pathways". These pathways, once isolated, can be rigorously characterized with classical psychophysical techniques: test and field action spectra, test and field mixture experiments, critical durations, adaptation dynamics, etc. The pathway through which the perturbation signal passes when a human observer's detection is governed by Stiles's π_1 and π_3 has been characterized by our lab and others with these techniques: Figure 10 schematically summarizes our knowledge. The pathway clearly comprises a stage of cone-antagonistic neural coding, and a site of gain control proximal to the cone-antagonistic coding. Much of the interesting adaptation dynamics of the pathway occurs at or proximal to the site of cone-antagonistic coding. Protanopes and deuteranopes appear to possess a similar pathway. The encoding of the signals that ultimately form the basis of blue/yellow hue equilibrium judgments in normals and dichromats appears to occur at the same cone-antagonistic site.

The characterization of a psychophysically-defined pathway such as π_1/π_3 can clearly provide rich insight into neural organization.

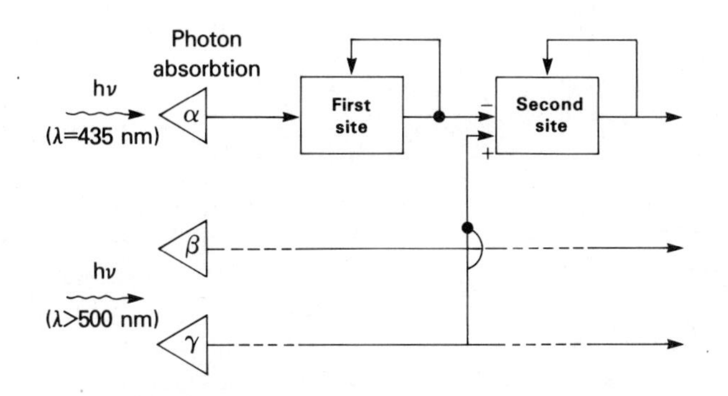

Figure 10.

Summary schematic of π_1/π_3 pathway.

Nonetheless, the danger of generalizing from a restricted set of conditions under which isolation is well established to others without checking applicable defining criteria of the pathway must be emphasized. Changing what may appear to be even minor aspects of the detection task (e.g., from a 2AFC a yes/no or adjustment procedure) may significantly change the nature of what is isolated. This is why in changing from the standard Stiles test flash (200 msec, 1 deg) to our low frequency test we have repeated the pathway-defining field sensitivity measurement (Figure 1) and directly compared field mixture curves with the two types of test targets (Figure 5). Sometimes what appears to be a simple change in conditions does not yield anticipated correspondence. Yim (1983) measured modulation thresholds for a short-wave stimulus under a variety of background conditions that isolated π_1 for simple incremental flashes, and showed that a stable low-bandwidth m.t.f. component curve (like that of Wisowaty & Boynton, 1980) could be measured whenever detection could be assigned to a modulation signal originating in the short-wave cones. The low-bandwidth m.t.f. component signal apparently traveled a pathway with a cone-antagonistic gain site, but did not have the same long-wave opponency as the cone-antagonistic gain effect studied by Pugh & Larimer (1980). We thus wish to make it clear that we do not believe that the pathway we have described here encompasses even all the signals that originate in the short-wave cones, and emphasize again that the theory proposed here applies explicitly only to a limited set of experimental conditions.

REFERENCES

Augenstein, E.J. & Pugh, Jr., E.N. (1977). The dynamic behavior of Stiles's π_1 colour mechanism: further evidence for two sites of adaptation. J. Physiol., 272, 247-281.

Friedman, L.J. (1983). Cone antagonism along visual pathways of red/green dichromats. Ph.D. Thesis, Univ. of Michigan.

Friedman, L.J., Thornton, J.E. & Pugh, Jr., E.N. (1984). Cone antagonism along visual pathways of red/green dichromats. Submitted.

Friedman, L.J., Yim, P. & Pugh, Jr., E.N. (1984). Temporal integration of the π_1/π_3 pathway in normal and dichromatic vision. Vision Res., in press.

Hurvich, L.M. & Jameson, D. (1957). An opponent process theory of color vision. Psychol. Rev., 64, 384-404.

Kranda, K. & King-Smith, P.E. (1979). Detection of colored stimuli by independent linear systems. Vision Res., 19, 733-745.

Krantz, D.H. (1975). Color measurement and color theory. I. Representation theorem for Grassmann structures. J. Math. Psych., 12, 283-303.

Mollon, J.D. & Polden, P.G. (1977a). An anomaly in the response of the eye to light of short wavelengths. Phil. Trans. Roy. Soc. Lond., 278, 207-240.

Mollon, J.D. & Polden, P.G. (1977b). Further anomalies of the blue mechanism. Invest. Ophthal. Vis. Sci., ARVO Supplement.

Pugh, Jr., E.N. (1976). The nature of the π_1 mechanism of W.S. Stiles. J. Physiol., 257, 713-747.

Pugh, Jr., E.N. & Larimer, J. (1980). Test of the identity of the site of blue/yellow hue cancellation and the site of chromatic antagonism in the π_1 pathway. Vision Res., 20, 779-788.

Pugh, Jr., E.N. & Mollon, J.D. (1979). A theory of the π_1 and π_3 color mechanisms of W.S. Stiles. Vision Res., 19, 293-312.

Sperling, H.G. & Harwerth, R.S. (1971). Red-green cone interactions in the increment threshold spectral sensitivity of primates. Science, 172, 180-184.

Stiles, W.S. (1939). The directional sensitivity of the retina and the spectral sensitivities of the rods and cones. Proc. Roy. Soc. B., 127, 64-105.

Stiles, W.S. (1949). Increment thresholds and the mechanisms of colour vision. Doc. Ophthal., 3, 138-163.

Stiles, W.S. (1953). Further studies of visual mechanisms by the two-colour threshold method. Coloq. Probl. Opt. Vis. (U.I.P.A.P., Madrid), 1, 65-103.

Stiles, W.S. (1959). Color vision: the approach through increment-threshold sensitivity. Proc. Natl. Acad. Sci., 45, 100-114.

Stiles, W.S. (1967). Mechanism concepts in colour theory. J. of the Colour Group, 105-123.

Stiles, W.S. (1978). Mechanisms of Colour Vision. Academic Press, New York.

Thornton, J.E. & Pugh, Jr., E.N. (1982). Threshold inhibition in the yellow/blue pathway. Invest. Ophthal. Vis. Scie., 22, 18, (ARVO Supplement).

Thornton, J.E. & Pugh, Jr., E.N. (1983a). Red/green color opponency at detection threshold. Science, 219, 191-193.

Thornton, J.E. & Pugh, Jr., E.N. (1983b). Relationships of opponent-colours cancellation measures to cone-antagonistic signals deduced from increment threshold data. In Color Vision: Physiology and Psychophysics. (eds. J.D. Mollon and L.T. Sharpe). Academic Press, New York. 361-373.

Thornton, J.E. & Pugh, Jr., E.N. (1984). Blue/yellow color opponency at increment threshold. In preparation.

Wisowaty, J.J. & Boynton, R.M. (1980). Temporal modulation sensitivity of the blue mechanism: measurements made without chromatic adaptation. Vision Res., 20, 895-909.

Wyszecki, G. & Stiles, W.S. (1982). Color Science (2nd edition). Wiley, New York.

Yim, P. (1983). The temporal characteristics of a short-wavelength sensitive pathway. Ph.D. Thesis, Univ. of Pennsylvania.

The Action Spectra of Rods and Red- and Green-Sensitive Cones of the Monkey *Macaca Fascicularis*

B.J. Nunn,[1] J.L. Schnapf[2] and D.A. Baylor[2]

[1]The Physiological Laboratory, University of Cambridge, Downing Street, Cambridge CB2 3EG, UK

[2]Department of Neurobiology, Stanford University School of Medicine, Stanford, Connecticut 94305, USA

INTRODUCTION

To understand colour vision in primates and man it is necessary to know the spectral sensitivities of the photoreceptors. However the action spectra of the cones cannot be determined directly from psychophysical measurements or from recordings from higher-order visual neurones. Microspectrophotometric measurements of the absorption spectra of the tiny amounts of visual pigments in single cones cannot be made reliably at absorptions less than about 1/10th of the maximum, so that these measurements are restricted to a limited range of wavelengths. We have measured the spectral sensitivities of single rods and cones from the monkey *Macaca fascicularis* over the whole visible spectrum using suction electrodes to measure the light-sensitive current. The scotopic spectral sensitivity and colour matches of this animal are similar to those of humans (Blough & Schrier, 1963; Morgan, 1966; DeValois *et al.*, 1974).

METHODS

Small pieces of retina, usually from the periphery, were obtained by mechanical dissociation. The outer segment of a rod or cone at the edge of a piece was drawn into a current-recording pipette. Light stimuli, plane polarized for optimal absorption, were applied at right angles to the photoreceptor axis. Measurements were made at a temperature near 37°C.

The sensitivity of each photoreceptor was determined at different wavelengths as the reciprocal of the flash intensity required to elicit a constant response. This method was used to determine the action spectra of cones in the turtle by Baylor and Hodgkin (1974) and is similar to the psychophysical "threshold versus radiance" technique employed by Stiles (1978). Frequent measurements were made at a reference wavelength to allow for slow changes in the condition of the cells.

RESULTS AND DISCUSSION

The responses recorded from the outer segments of rods and cones of *Macaca fascicularis* consisted of outward-going currents with saturating amplitudes of up to 35 pA in the rods and 20 pA in the cones. The rod responses were slower and more sensitive than those of the cones. The peak response to a dim flash usually occurred after about 150-250 msec in rods and after 60-100 msec in cones. Both the rods and the cones obeyed the principle of univariance (Rushton, 1972) in that the kinetics of the dim-flash response and the form of the response-intensity relation were the same at different wavelengths.

Spectral sensitivities

Action spectra of 10 rods and 7 cones were determined over the visible spectrum at intervals of 20 nm. The action spectra of the rods had a peak in the blue-green region of the spectrum, while the action spectra of the cones fell into two groups with maxima in the green and yellow-green (Figs 1-3). Following the usual convention the two classes of cones are termed 'green-sensitive' (Fig. 2) and 'red-sensitive' (Fig. 3). We have not yet recorded from 'blue-sensitive' cones.

It was of interest to compare the action spectra of monkey photoreceptors with the estimated action spectra of human rods and cones. This comparison is easiest for rods since human scotopic vision is mediated by only one kind of photoreceptor. Fig. 1 shows that at wavelengths longer than about 600 nm the form of the average action spectrum of the monkey rods (circles) is very similar to the scotopic visibility curve determined by Crawford (1949) for human observers using a brightness-matching method (filled triangles). The open triangles in Fig. 1 show that Crawford's (1949) curve also agreed with the action spectrum of the monkey rods at shorter wavelengths when his curve was corrected for lens absorption and self-screening of rhodopsin.

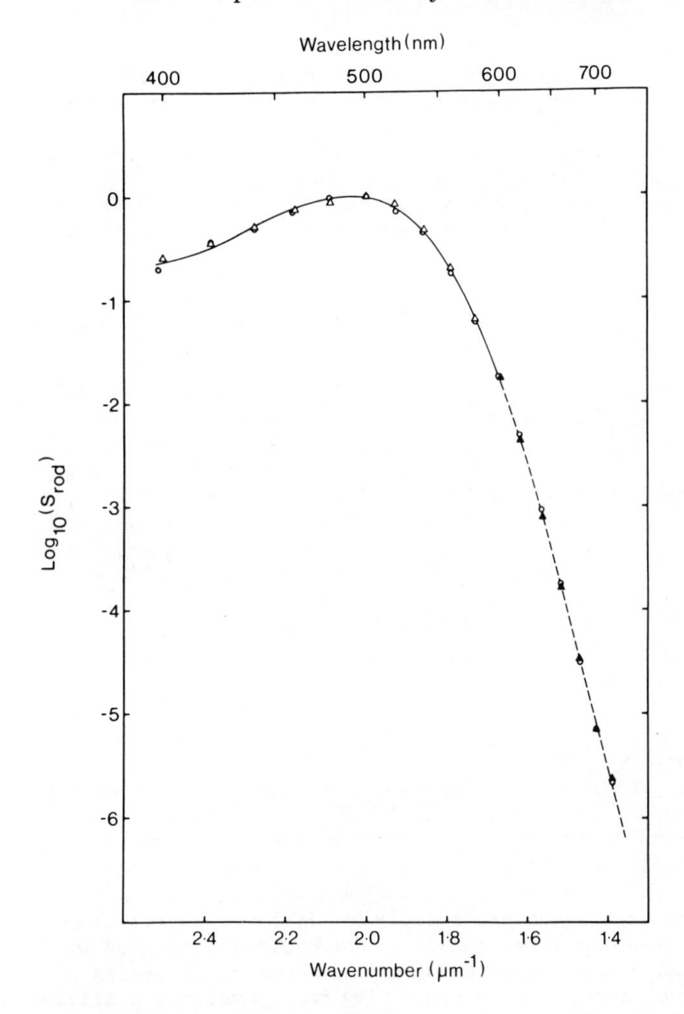

Fig. 1. Average spectral sensitivity of 10 rods from 3 monkeys (circles) compared with action spectrum estimated for human rods (triangles) and the Dartnall nomogram for a rhodopsin having a peak absorption at 491 nm (smooth curve, from Wyszecki & Stiles, 1982). Interrupted curve is arbitrary. Filled triangles show scotopic visibility curve measured by Crawford (1949) for $\lambda \geq 600$ nm. Open triangles are Crawford's (1949) curve at shorter wavelengths, corrected for rhodopsin self-screening and for lens absorption. The form of the lens optical density spectrum used in the correction was that tabulated by Wyszecki and Stiles (1982); the lens optical density was taken as 0.17 at 500 nm. The correction for self-screening assumed an axial rhodopsin density of 0.36 at 500 nm.

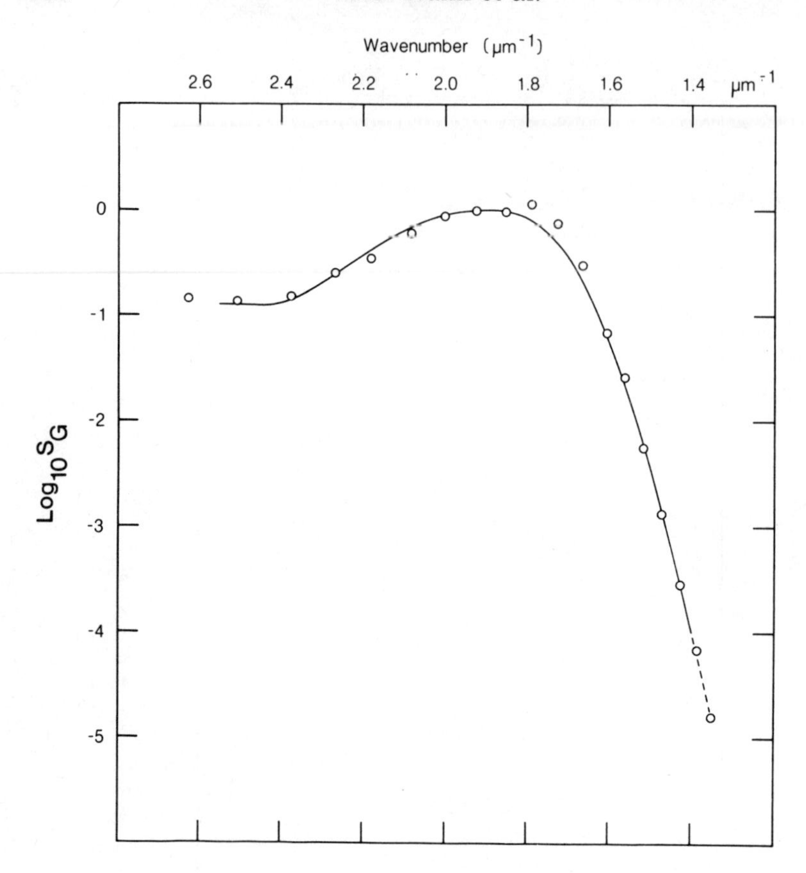

Fig. 2. Average spectral sensitivity of 3 peripheral green-sensitive monkey cones (points) compared with the average action spectrum estimated for cones of the same class in man by Estevez (Wyszecki & Stiles, 1982). Psychophysical results corrected for absorption by the lens and macular pigment and for self-screening by the visual pigment (see Text).

The action spectra of human cones cannot be derived directly from colour-matching results without making assumptions. A common assumption follows the proposal of König (1903) that dichromats simply lack one of the cone visual pigments of normal vision. In that case the action spectra of the three cone mechanisms can be derived from colour matches of normal trichromats and from the location in the trichromatic chromaticity diagram of the 3 confusion points of the 3 kinds of dichromat. The continuous curves in Figs. 2 and 3 are the

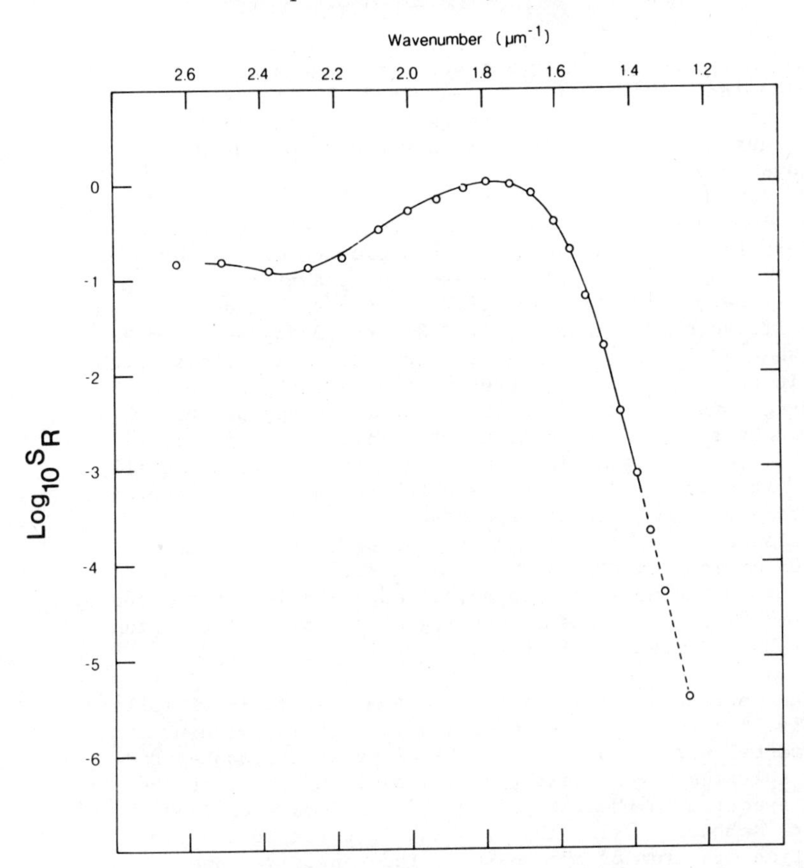

Fig. 3. Average spectral sensitivity curve of one foveal and three peripheral red-sensitive cones from the monkey (points) compared with the curve estimated by Estevez (Wyszecki & Stiles, 1982) for human red-sensitive cones; curve corrected for optical absorption as in Fig. 2.

action spectra estimated in this way by Estevez (Wyszecki & Stiles, 1982) after correction for optical densities of the lens of 0.125 at 500 nm, of the macular pigment of 0.46 at 460 nm and for peak optical densities of the visual pigments of 0.3. Estevez used the data of Stiles and Burch (1955) for colour matches of normal trichromats, and obtained average dichromatic confusion points from several different studies. The action spectra derived for human green-and red-sensitive cones agree reasonably with the average action spectra measured for the monkey cones, suggesting that the action spectra of the green- and red-sensitive cones in man and Macaca fascicularis are similar. The fact that Estevez's average dichromatic confusion loci* (Wyszecki & Stiles, 1982) led to the good agreement in Figs. 2 and 3 supports the proposal of König (1903) that dichromatic vision is a reduced form of normal trichromatic vision.

At long wavelengths the action spectra of the monkey receptors fell on straight lines when plotted on a wavenumber scale. The average slopes of these lines, expressed as log units.μm, were 15.0, 15.5 and 17.7 for the rods and green- and red-sensitive cones respectively. Stiles (1948) advanced a model that explains why the action spectra fall along straight lines in the far red. He suggested that the absorption of low-energy photons depends on the thermal energy of the absorbing visual pigment molecule to provide the additional energy needed for isomerization. Then, for numerous, equally spaced vibrational energy levels which are populated according to Boltzmann's Law, the absorption spectrum should be proportional to $e^{-(hc/\lambda kT)}$ and the slope of the action spectrum at long wavelengths should be 20.1 μm, which is steeper than the measured slopes. A modification introduced by Lewis (1955) predicts the steepening of the slope for action spectra having peaks at longer wavelengths.

The corresponding slope for the human scotopic visibility curve is 14.8 μm (from Crawford, 1949) which is, within experimental error, the same as the slope of the monkey rod action spectrum. For human foveal vision the slope of the action spectrum in the far red is 17.4 μm (Goodeve, 1936; Griffin, Hubbard & Wald, 1947), which is close to the slope of the action spectrum of the monkey's red-sensitive cones.

Form of the spectral sensitive curves

The wavelength of peak sensitivity of the monkey rods was estimated by comparing the average action spectra with the Dartnall nomogram for retinal_1-based visual pigments (Wyszecki & Stiles, 1982). When plotted on a wavenumber scale the average rod action spectrum fitted the nomogram with a maximum absorption at 491 nm (Fig. 1). However the action spectra of the cones were significantly narrower than the Dartnall nomogram on the $1/\lambda$ scale. Barlow (1982) showed that the absorption spectra measured from single human photoreceptor

*Estevez (private communication) has pointed out that the values of r_{pc} and g_{pc} in Wyszecki and Stiles (1982) were printed incorrectly and ought to be 1.0381 and 0.0388 respectively; the tabulated action spectra were printed correctly.

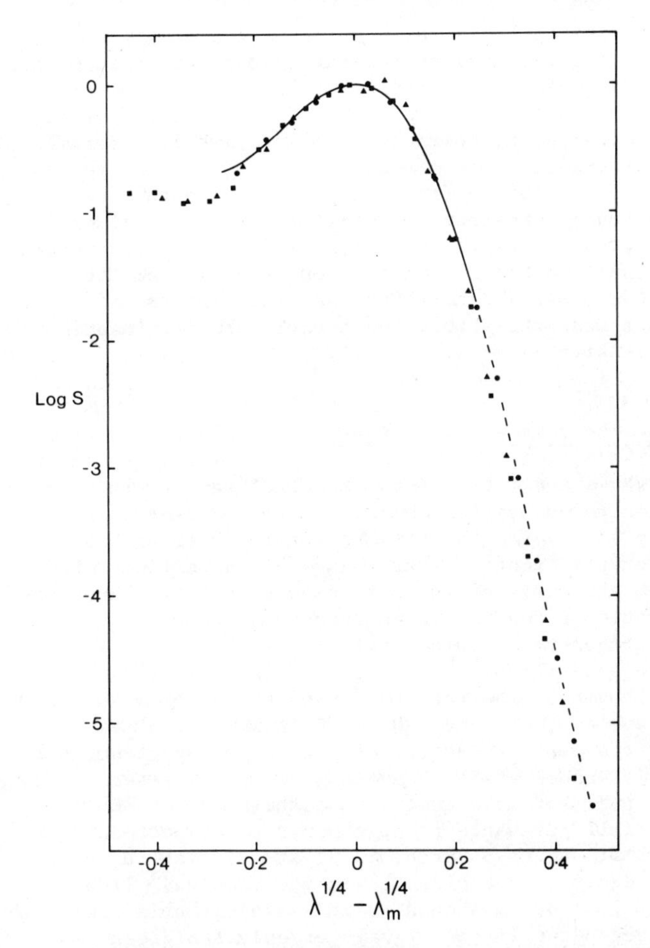

Fig. 4. Comparison of action spectra of monkey rods and cones with Dartnall nomogram (λ_m = 500 nm, smooth curve) on scale of $\lambda^{1/4}$. Interrupted curve has been drawn through rod action spectrum by eye. Values of λ_m used were: rods, ●, 491 nm; green-sensitive cones, ▲, 533 nm; red-sensitive cones, ■, 565 nm.

outer segments by Bowmaker, Dartnall and Mollon (1980) were very similar on a scale of $\lambda^{1/4}$. Fig. 4 show the average action spectra of the monkey rods and cones fitted to the Dartnall nomogram on this scale. The wavelengths of maximum sensitivity used in this plot were 491 nm for the rods and 533 nm and 565 nm for the green- and red-sensitive cones respectively. The curves in Fig. 4 all have similar shapes but the cone action spectra are slightly narrower than the rod action spectrum and the Dartnall nomogram.

The values of λ_m estimated for the monkey rods and cones are similar to the wavelengths of peak absorption measured for the corresponding photoreceptors by microspectrophotometry in *Macaca fascicularis* (Bowmaker *et al.*, 1980; MacNichol *et al.*, 1983) and man (Dartnall, Bowmaker & Mollon, 1983). Dartnall *et al.* (1983) determined average values of λ_m of 496 nm from human rods and 531 nm and 558 nm from the long-wavelength cones of normal trichromats. Despite the similarity of the values of λ_m for the absorption and action spectra, the absorption spectra were significantly broader than the corresponding action spectra. It is possible that light scattered by the photoreceptors made the absorption spectra broader (MacNichol *et al.*, 1983; Dartnall *et al.*, 1983).

The ratios of the cone sensitivities

Fig. 5 shows the ratio of the spectral sensitivities of the green- and red-sensitive cones. The ratio decreases monotonically at wavelengths between about 460 nm and 700 nm so that monochromatic lights in this range could be identified uniquely from the ratio of the responses of the two kinds of cone. The ratio of sensitivities increases again at wavelengths longer than 700 nm; this means that light at these wavelengths will be indistinguishable from light at shorter wavelengths. The perceived "yellowing" of light of increasing wavelength beyond 700 nm (Brindley, 1955) is explained by the gentler descent of the action spectrum of the green-sensitive cones at long wavelengths as suggested by Lewis (1955). The pairs of wavelengths on either side of 700 nm which are indistinguishable in psychophysical experiments (Brindley, 1955; Stiles & Burch, 1959) are similar to those which can be derived from Fig. 5. In the same way, light at wavelengths below 460 nm should be indistinguishable from light at longer wavelengths if the signals of only the green- and red-sensitive cones are considered. This yellowing of light at wavelengths less than 460 nm has been observed in tritanopes by Wright (1952) and in normal trichromats for very small foveal stimuli, which result in tritanopic behaviour, by Willmer and Wright (1945). Reasoning along these lines also suggests that wavelength discrimination would be poorest at the maxima and minima of the curve in Fig. 5. Indeed, discrimination by tritanopes and by normal trichromats is particularly poor for small foveal stimuli at wavelengths around 460 nm (Willmer & Wright, 1945; Wright, 1952).

The relatively small separation of about 30 nm between the peaks of the action spectra of the green- and red-sensitive cones has been considered somewhat surprising (e.g. Barlow, 1982), but may reflect a compromise aimed at sacrificing fine

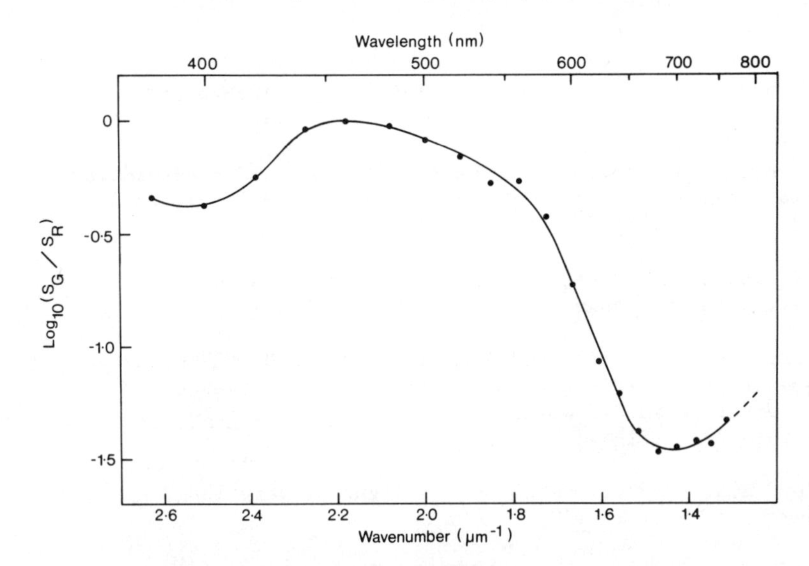

Fig. 5. Log_{10} of ratio of sensitivities of green- and red-sensitive cones. Curve drawn by eye; interrupted portion extrapolated assuming spectra in Figs. 2 & 3 continue along straight lines in the far red.

spectral resolution for the ability to perceive colours over a wide range of wavelengths. Suppose that the action spectrum of the red-sensitive cones were displaced to a peak at 650 nm, and that the action spectrum maintains its shape on the $\lambda^{1/4}$ scale. In this case the ratio of sensitivities of the green- and red-sensitive cones would have a maximum near 510 nm and would fall very steeply at longer wavelengths. If the ratio of sensitivities is restricted to a useable range of 1.5 log units (as in Fig. 5), then unique colour-naming mediated by the two long-wavelength mechanisms would be restricted to wavelengths between 510-610 nm, which is less than half the usual range. Wavelength resolution would be reduced if the separation between the action spectra of the green- and red-sensitive cones were made smaller (as may be the case for some anomalous trichromats), since for a vanishingly small separation the ratio of sensitivities of the cones would not vary with wavelength.

We thank Mr. Mark Siegel and Drs Carolyn Reed, Marla Luskin, Helga Kolb, Walter Makous and Carla Shatz for valuable assistance and Drs Peter McNaughton and Trevor Lamb for helpful comments on this manuscript. Prof. Horace Barlow pointed out the similarity of the cone action spectra on the $\lambda^{1/4}$ scale. This work was supported by US PHS grant EY01543.

REFERENCES

Barlow, H.B. (1982). What causes trichromacy? A theoretical analysis using comb-filtered spectra. Vision Res. 22, 635-643.

Baylor, D.A. and Hodgkin, A.L. (1973). Detection and resolution of visual stimuli by turtle photoreceptors. J. Physiol. 234, 163-198.

Blough, D.S. and Schrier, A.M. (1963). Scotopic spectral sensitivity in the monkey. Science 149, 493-494.

Bowmaker, J.K., Dartnall, H.J.A. and Mollon, J.D. (1980). Microspectrophotometric demonstration of four classes of photoreceptor in an Old World primate Macaca fascicularis. J. Physiol. 298, 131-143.

Brindley, G.S. (1955). The colour of light of very long wavelength. J Physiol. 130, 35-44.

Crawford, B.H. (1949). The scotopic visibility function. Proc. Phys. Soc. B62, 321-334.

Dartnall, H.J.A., Bowmaker, J.K. and Mollon, J.D. (1983). Human visual pigments: microspectrophotometric results from the eyes of seven persons. Proc. Roy. Soc. B220, 115-130.

DeValois, R.L., Morgan, H.C., Polson, M.C., Mead, W.R. and Hull, E.M. (1974). Psychophysical studies of monkey vision. I. Macaque luminosity and colour vision tests. Vision Res. 14, 53-67.

Goodeve, C.F. (1936). Relative luminosity in the extreme red. Proc. Roy. Soc. Lond. A155, 664-683.

Griffin, D.R., Hubbard, R. and Wald, G. (1947). The sensitivity of the human eye to infra-red radiation. J. Opt. Soc. Am. 37, 546-554.

König, A. (1903). Gesammelte Abhandlungen. Barth-Verlag, Leipzig. From Wyszecki and Stiles, 1982.

Lewis, P.R. (1955). A theoretical interpretation of spectral sensitivity curves at long wavelengths. J. Physiol. 130, 45-52.

MacNichol, E.F., Levine, J.S., Mansfield, R.J.W., Lipetz, L.E. and Collins, B.A. (1982). Microspectrophotometry of visual pigments in primate photoreceptors. In Colour Vision. (eds. J.D. Mollon and L.T. Sharpe). Academic, London.

Morgan, A.A. (1966). Chromatic adaptation in the macaque. J. Comp. Physiol. Psychol. 62, 76-83.

Rushton, W.A.H. (1972). Pigments and signals in colour vision. J. Physiol. 220, 1-31P.

Stiles, W.S. (1948). The physical interpretation of the spectral sensitivity curve of the eye. In Transactions of the Optical Convention of the Worshipful Company of Spectacle Makers, pp. 97-107. London: Spectacle Maker's Company. Reprinted in Stiles (1978).

Stiles, W.S. (1978). Mechanisms of Colour Vision. Academic, London.

Stiles, W.S. and Burch, J.M. (1955). Interim report to the Commission Internationale de l'Eclairage, Zurich, 1955, on the National Physical Laboratory's investigation of colour-matching. Optica Acta 2, 168.

Stiles, W.S. and Burch, J.M. (1959). N.P.L. colour-matching investigation: final report (1958). Optica Acta 6, 1-26.

Willmer, E.N. and Wright, W.D. (1945). Colour sensitivity of the fovea centralis. Nature, 156, 119-121.

Wright, W.D. (1952). Characteristics of tritanopia. J. Opt. Soc. Am. 42, 509-521.

Wyszecki, G. and Stiles, W.S. (1982). Colour Science. Wiley, New York.

Properties of Cone Photoreceptors in Relation to Colour Vision

T.D. Lamb

Physiological Laboratory, Downing Street, Cambridge CB2 3EG, UK

INTRODUCTION

The chapter by Nunn, Schnapf & Baylor (this volume) describes how primate photoreceptors respond to illumination. I shall be continuing with this theme, describing further important properties of the cone photoreceptors, and discussing the implications of these properties for colour vision. In particular there is one important property of cone photoreceptors, namely their "automatic gain control", which has profound implications for colour vision, and I shall be expanding in some detail on that. These implications are also mentioned in the chapter by MacLeod (this volume).

The title of this session of the meeting is "the problem of colour constancy". One of the very important characteristics of our sense of colour vision is that our subjective sensation of the colours of reflecting objects is only weakly dependent on the ambient level of illumination. If we observe a visual scene, for example a vase of flowers, then our overall impression of the colours is not very greatly affected by whether we are indoors or outdoors, whether it is a very dull day or a bright sunny day. That is not to say that the colours do not appear subjectively a little more intense in bright sunlight, but qualitatively they are obviously the same colours and we give them the same names. So for reflected scenes we have "colour constancy" or "hue constancy" over a range of intensities of at least a million-fold, from less than 1 troland in a dimly-lit room to around a million trolands on a bright sunny beach.

We perceive constancy of colour not only when the intensity of illumination changes, but also (as shown in the demonstration by Edwin Land) when the wavelength compositon of the illumination is changed over a remarkably wide range.

So, more-or-less irrespective of either the intensity or the colour composition of the illumination, we subjectively perceive the colour of reflected objects as invariant. How can this be possible? The answer I think must take account of the photoreceptors - and in particular of their automatic gain control.

CONE ADAPTATION - TURTLE

Unfortunately the appropriate experiments have not yet been performed using recordings from single primate cones. So I shall be relying on experiments on the cones of lower vertebrates, but we have no reason to suspect any major differences - indeed a number of experiments indicate that primate cones and lower cones are remarkably similar.

Basically, what is found is that cones adapt in a Weber-law manner. Over a wide range of intensities the incremental sensitivity of a cone is inversely proportional to the background intensity. In other words steady illumination desensitizes the cone and the desensitization is proportional to the intensity of the steady illumination.

This is shown in a turtle cone in Fig. 1 (Baylor & Hodgkin, 1974), which illustrates intracellular voltage responses to brief flashes of light. The trace labelled 1 is the average response to

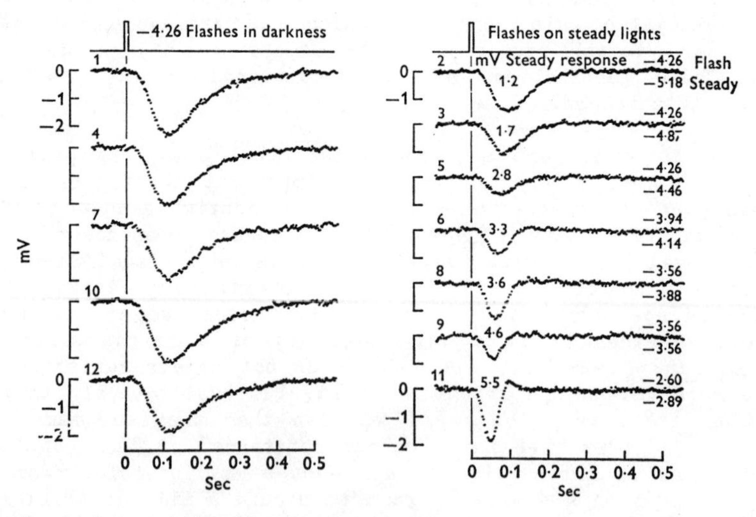

Fig. 1. Desensitization of turtle cone by steady lights. Left-hand column shows control responses to flashes presented in darkness; right-hand column shows responses in the presence of backgrounds of increasing intensity. Numbers at right represent flash intensity and steady intensity in $\log_{10}$ units. Reproduced with permission from Baylor & Hodgkin (1974).

a dim flash presented in darkness, and has a time to peak of about 100 msec and, for this intensity, has an amplitude of 2 mV. Trace 2 on the right is the response to the same flash presented on a dim steady background, and is slightly smaller in amplitude. In trace 3 the background is slightly brighter, and the response is smaller again. Trace 4 is a control in darkness to show that the cell has not deteriorated, and trace 5, with a brighter background, shows the response to have declined even further. Then as we continue down the right hand side the steady background was made brighter each time, but after the top three traces the test flashes were also made brighter, simply in order that a measurable response could be obtained. And it is clear that in the presence of these desensitizing backgrounds the flash response gets faster - with the brightest background the time to peak is about 50 msec, whereas in darkness it is a little greater than 100 msec.

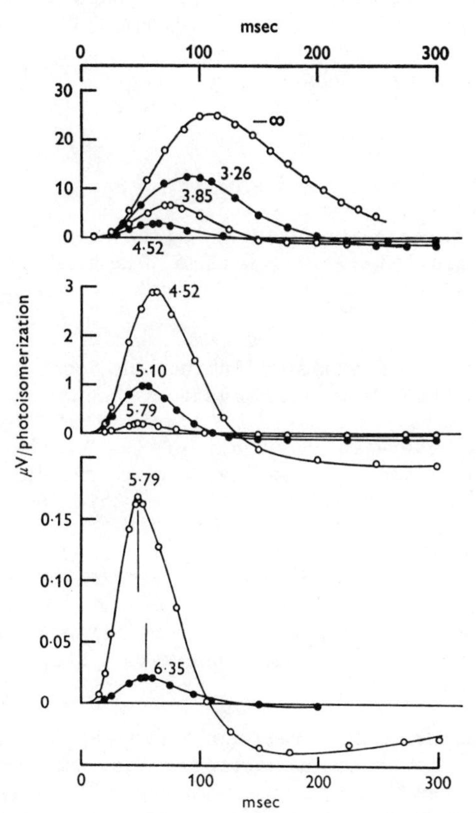

Fig. 2. Desensitization of turtle cone by steady lights. Superimposed responses to flashes in the presence of backgrounds of increasing intensity. Numbers near traces represent $\log_{10}$ of background intensity expressed in photoisomerizations sec^{-1} $cone^{-1}$. Reproduced with permission from Baylor & Hodgkin (1974).

Similar results are replotted in Fig. 2. At the very top is the response in the absence of a background (log background $-\infty$), and below are superimposed responses in the presence of backgrounds of increasing intensity. Generally speaking the responses begin rising along a common curve, but then depart from it at earlier and earlier times. In the lower panels are magnified plots showing what happens at higher intensities.

The ordinate in this Figure is μV/photoisomerizations - that is, sensitivity. One can take the peak response, or sensitivity, and plot this as a function of steady background intensity as shown in Fig. 3. In fact reciprocal sensitivity ($1/S_F$) is plotted here - that is the flash intensity required to elicit a criterion response amplitude. In psychophysical terms this is the equivalent of a threshold-versus-intensity curve, and it shows Weber-law behaviour: threshold is proportional to background plus a constant (this plot is in linear co-ordinates). So we have

$$1/S_F = 1/S_F^D + c\,I\,, \qquad (1)$$

equivalent to

$$\Delta I = \Delta I_0 + k\,I\,. \qquad (2)$$

To test this equation over a wider range of intensities $1/S_F - 1/S_F^D$ was plotted (Fig. 4) as a function of intensity in double log co-ordinates, and the straight line with unity slope shows Weber's law.

So in turtle cones steady illumination desensitizes the cone and improves its time response, just as occurs in the overall visual system. And it seems quite likely that for the overall visual system the Weber-law behaviour of the photopic system with large area stimuli, and the improved time resolution at higher photopic levels, may simply result from this behaviour of the photoreceptors.

As far as the physical basis of this "automatic gain control" is concerned, it appears to be internal to the transduction mechanism in the cone's outer segment, and presumably results from alterations in the biochemical steps which lead to the electrical response.

For the present considerations, however, I am not so much concerned with how the cone properties arise, but rather with the question of how these properties influence colour vision. Before looking at that question though, let me try to compare the data on turtle cones with the data on human psychophysics.

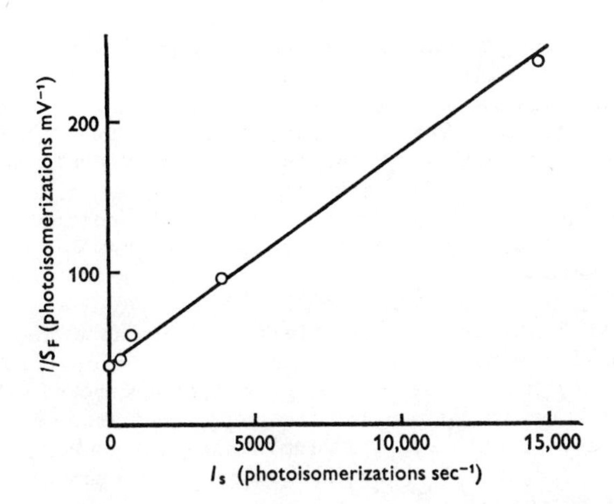

Fig. 3. Relation between reciprocal of flash sensitivity and steady light intensity, with both coordinates linear. From same cell as Fig. 2. Reproduced with permission from Baylor & Hodgkin (1974).

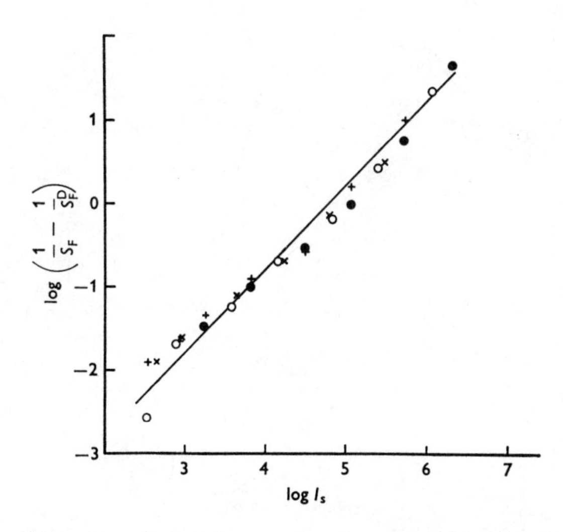

Fig. 4. Effect of background on sensitivity. Note that by re-arranging eqn. (1) Weber's law predicts a straight line of unity slope in double logarithmic co-ordinates. Data from three cells. Reproduced with permission from Baylor & Hodgkin (1974).

COMPARISON WITH PSYCHOPHYSICS

Data from human psychophysics is shown in Fig. 5. This is from the work of Kelly (1961) and shows the threshold for detection of flicker in a large test field which is modulated sinusoidally in time. The abscissa plots the frequency of modulation, and the ordinate plots the depth of modulation required for flicker to be detectable. At a dim level of illumination (0.65 trolands) flicker is most noticeable at a frequency of 5 Hz, and at higher frequencies the sensitivity to flicker rapidly falls.

As the mean luminance of the field is increased the observer becomes more sensitive to rapid flicker. The frequency of maximum flicker sensitivity increases and the modulation required at the peak declines. This is shown by the fact that the maxima move upward and to the right as the luminance increases. And the "critical fusion frequency" (the frequency corresponding to 100% modulation) also increases.

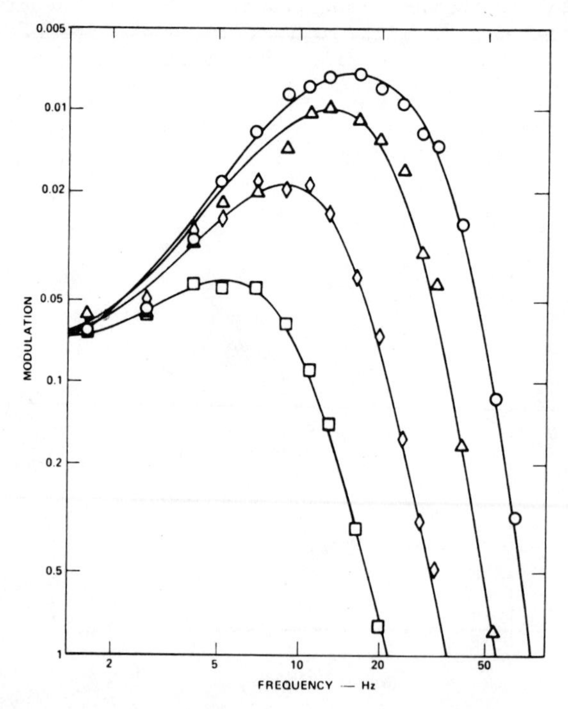

Fig. 5. Flicker threshold in a human observer. The depth of sinusoidal modulation required for threshold perception of flicker in a large field is plotted as a function of frequency of modulation, at four background luminances (from below: 0.65, 7.1, 77, and 850 trolands). Reproduced with permission from Kelly (1971), after Kelly (1961).

These data can be replotted in terms of the absolute amplitude of modulation, as shown in Fig. 6. So instead of percentage modulation this figure plots modulation in trolands - that is, percentage modulation times mean luminance. And the behaviour shown in the Figure is exactly what is expected for a detector which desensitizes at low frequencies in a Weber-law manner and which speeds up in the way that the cone is known to. In particular, although there is Weber-law behaviour at low frequencies, at high frequencies the curves all asymptote to a common curve. This corresponds broadly to the observation that the cone flash responses all rise along a common curve at early times, independent of background (see Fig. 2).

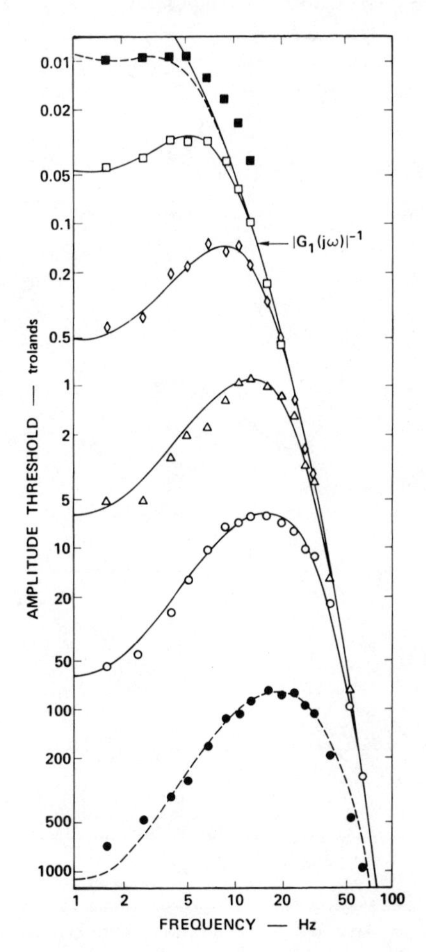

Fig. 6. Absolute amplitude of modulation for the perception of flicker. Some data as Fig. 5, with the addition of one dimmer and one brighter level of mean luminance. Reproduced with permission from Kelly (1971), after Kelly (1961).

From these data Kelly (1971) attempted to calculate back to the form of the underlying impulse response, but this cannot be done uniquely in the absence of phase information. Nevertheless by making certain assumptions he obtained the following impulse responses (Fig. 7). The top trace is for the lowest luminance, and in each of the lower traces the luminance increases by about a factor of 10. Obviously these curves bear quite a striking similarity to the turtle cone responses, speeding up as the intensity increases.

This result suggests that the adaptational behaviour of the responses of turtle cones and of human cones may be very similar, and the following analysis will be based on this assumption.

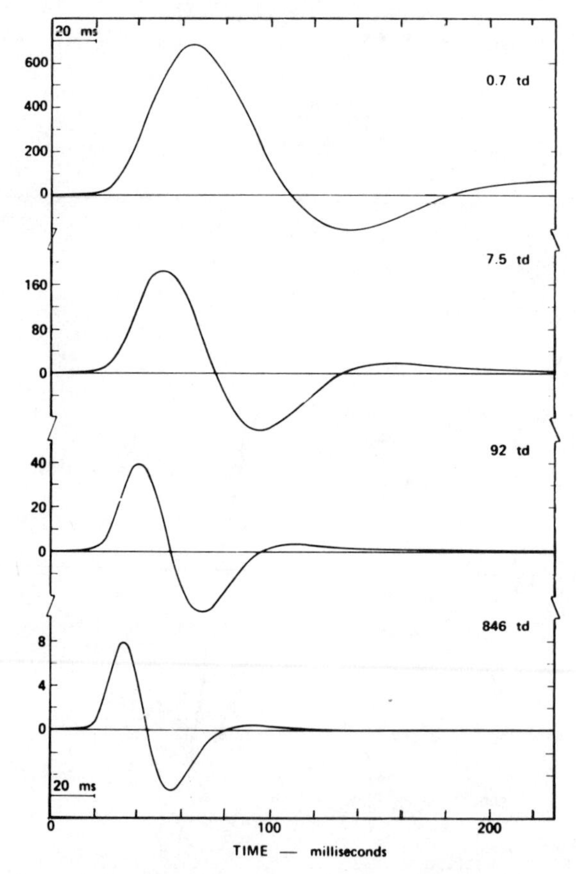

Fig. 7. Computed time course of responses which would generate the curves in Figs. 5 and 6. Note that the computed responses are not unique because of the absence of phase information. There is, however, a striking qualitative similarity with the responses in turtle cones (Fig. 2). Reproduced with permission from Kelly (1971).

RELEVANCE TO COLOUR VISION

The question now is: If primate cones do indeed desensitize in a Weber-law manner, what implications has this for colour vision?

As has recently been so elegantly pointed out by Shapley & Enroth-Cugell (1984), such Weber-law behaviour implies that the cones signal contrast. In other words, at moderate to high intensities, a fixed contrast $\Delta I/I$ elicits a fixed response amplitude ΔV, given by

$$\Delta V = \Delta I/I \; S_F^D \; I_0 \,, \tag{3}$$

where S_F^D is the sensitivity in darkness and I_0 is the intensity at which the sensitivity is halved.

Shapley & Enroth-Cugell maintain (and I fully agree) that this ability to signal contrast irrespective of steady working level is probably of immense importance to the visual system. Contrast is important to us because we view the real world as a scene in reflected light, and in a reflected scene contrast is invariant with the level of illumination. Our brain presumably wants to be presented primarily with information about the scene rather than with information about the ambient light level. And similar arguments apply to the case of colour constancy.

In order to consider the effect of cone desensitization on colour vision let us do the following. Let us imagine a region of the visual field to be uniformly illuminated with a steady intensity coloured light - and let us initially assume this to be monochromatic, of some wavelength λ. Let us next assume that the intensity of this uniform background is incremented, or alternatively that it is modulated (say sinusoidally). And let us finally assume that we are indeed in the Weber-law regions.

Then if we consider the red-sensitive (R) cones they will give some arbitrary signal, according to the contrast of the increment. And if we consider the green-sensitive (G) cones, then they will give exactly the same signal, because the contrast of the increment is the same to them. So irrespective of the wavelength of this field an increment or modulation of the same wavelength will generate an identical signal in each class of cones. Put another way, if this wavelength is near the peak absorption for one cone class then it will greatly desensitize that cone class, whereas if it is only weakly absorbed then it will only weakly desensitize the cones.

I should say that this precise equivalence depends on the so-called "Weber fraction" being the same for each class of cone. This is quite likely to be approximately true, but even if it is not this does not significantly change the argument. Because, put in different terms, modulation of the field generates a characteristic response in each class of cones which is independent of the wavelength composition of the field.

So you can imagine another separate uniform field of a different wavelength μ, being modulated about its mean luminance. Then, if the modulation is the same in the two cases, the signals in the various cone classes must be identical in the two fields. This, I emphasize, is for the case of modulation having the same spectral composition as the background - that is, we are adding or subtracting the same wavelength, not a different wavelength.

So far I have not mentioned differences in timing. If one class of cones is only weakly adapted then its responses will presumably be somewhat slower than those of another class of cones which is strongly adapted. However, this effect will probably not be very important because it is unusual for the stimulation of R and G cones to differ by much more than a log unit, because their absorption spectra are so close together. In any case, at high luminances the responses stop getting faster - they reach a limiting speed and only get less-sensitive - and in this region the time course of responses would be independent of wavelength. So I do not think that the possibility of different time-courses is an escape route from this apparent dilemma.

Where does this leave us?

Firstly, it shows that it is not possible to determine the colour of a modulated or flickering background at a fixed retinal position simply by comparing the responses of the different classes of cones in that one area, because they are all giving the same output. Nor is it possible to determine the colour by comparing one field with any other, if the fields are fixed on the retina, because the signals generated in the different areas are independent of the field colour.

In psychophysical terms this would predict that the "colour" of a stabilised image (that is the colour of an image perfectly stabilised on the retina) should disappear. And indeed this is known to be the case. When an image _is_ properly stabilised one very rapidly loses all perception of colour. And in fact, psychophysically, colour disappears more rapidly than does perception of luminance or of form.

I would further predict that if regions like these in a stabilised image were flickered or modulated relatively slowly then perceptually one should be aware of the luminance signal (i.e. aware of the presence of a flickering field and of its form), but unaware of the _colour_ of the region. I do not know whether this experiment has been performed, but I would be very interested to know the result.

My main point is simply that eye movements must be of fundamental importance to colour vision and that any theory of colour vision which ignores eye movements and object movements and which simply considers static stimulation cannot be correct.

Furthermore, there are implications for physiological experiments. If an animal is immobilised, and a stimulus is presented at a fixed position on its retina, and simply modulated in intensity, then the adaptational effects I have discussed will be critically important.

RECEPTOR COUPLING AND NOISE

At this stage I shall change the subject completely and discuss signal-to-noise ratio in cones, and the effects of electrical coupling between cones. Up until now I have considered only the *signals*, but it turns out that there is a large amount of *noise* in the cones in the presence of which the signals have to be detected.

Fig. 8. shows voltage recordings from two cones in the retina of the turtle (Simon, Lamb & Hodgkin, 1975). Below is what we would call a "typical" cone. In darkness there was a small amount of noise, and during a bright step of illumination the cell hyperpolarized as expected and the noise level was reduced - the trace in bright light is thinner than the trace in darkness before and after.

From the size of the receptive field of this cone (which covered perhaps 100 μm diameter) we believe the cell to have been strongly electrically coupled to neighbouring cones. In both rods and cones there is very good anatomical and physiological evidence for the existence of such electrical coupling, mediated by gap junctions between the receptors. Because of this electrical coupling signals spread over considerable distances (of the order of 5-10 cell diameters) across the retina, so that the recorded voltage in one cell is actually the average response of the light falling on a number of cells. As a result of this coupling any noise generated in the cells will be averaged out, because the noise in adjacent cells will be uncorrelated. Hence the noise level in this cell will be much smaller than would have been observed in the absence of electrical coupling.

Above is a rare example of what we believe to have been an "electrically isolated" cone - that is, a cone which for some reason was not electrically coupled to its neighbours. So this cone was displaying its full intrinsic dark noise, and that noise had a peak-to-peak level of about 3 mV (3000 μV). For comparison the mean response to a single photoisomerization per cone is about 30 μV (Baylor & Hodgkin, 1973), so that the single photon response would be completely undetectable beneath this noise, and it would require the simultaneous arrival of perhaps 100 or so photons to give a reliably detectable response (Lamb & Simon, 1977).

The vast majority of cells, however, are electrically coupled in peripheral regions of the retina, so that this situation is unusual. The effect of the coupling is to reduce the noise in the cells, and thereby to improve the signal-to-noise ratio, at the expense of reducing spatial resolution. In the

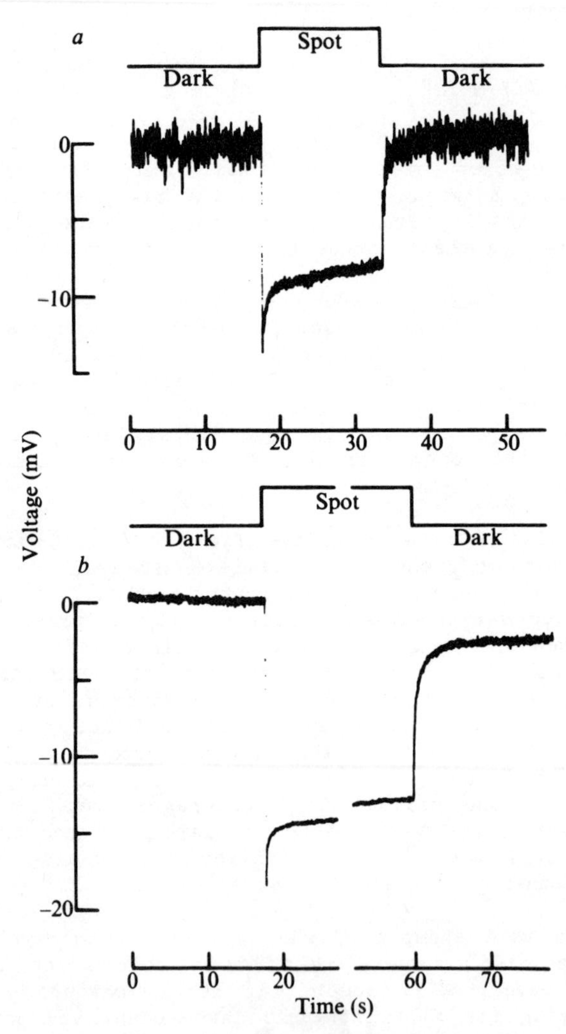

Fig. 8. Dark noise in two turtle cones, and its suppression by bright light. A, from a cone presumed to be electrically isolated from its neighbours; B, from an electrically coupled cone. Reproduced with permission from Simon, Lamb & Hodgkin (1975).

peripheral retina it is known that spatial resolution is nowhere near what is theoretically possible from the measured spacing between receptors, and it seems likely that spatial resolution at the level of the photoreceptors has been sacrificed for improved signal-to-noise ratio (Lamb & Simon, 1976). In this way the retina may be able to economise on synaptic connections, because it is no longer necessary for bipolar cells to contact every photoreceptor. In addition, it can be shown that there are advantages from the point of view of the signal-to-noise ratio of synaptic transmission, as discussed by Falk & Fatt (1972).

It has been shown by Detwiler & Hodgkin (1979) that in the turtle retina this coupling is very specific, so that red-sensitive cones are coupled only to other red-sensitive cones and green-sensitive cones are coupled only to other green-sensitive cones. This is precisely what one would expect if colour information is not to be thrown away at the outset.

In the human perifoveal and peripheral retina I would expect it quite likely that such electrical coupling would exist between R cones and between G cones, and I would expect coupling almost certainly to exist between human rods.

SUMMARY

In summary, it seems to me that in trying to comprehend our sense of colour vision it is necessary to take full account of the properties of the cone photoreceptors. Not only must we consider the absorption properties of their photopigments, but also the full range of properties of their electrical response to light, including temporal properties, adaptational properties, and also signal-to-noise considerations.

ACKNOWLEDGEMENT

The author was supported by grants from the Royal Society and the M.R.C.

REFERENCES

Baylor, D.A. and Hodgkin, A.L. (1973). Detection and resolution of visual stimuli by turtle photoreceptors. J. Physiol. 234, 163-198.

Baylor, D.A. and Hodgkin, A.L. (1974). Changes in time scale and sensitivity in turtle photoreceptors. J. Physiol. 242, 729-758.

Detwiler, P.B. and Hodgkin, A.L. (1979). Electrical coupling between cones in turtle retina. J. Physiol 291, 75-100.

Falk, G. and Fatt, P. (1972). Physical changes induced by light in the rod outer segment of vertebrates. In Handbook of Sensory Physiology. VII/1. Photochemistry of Vision. (ed. H.J.A. Dartnall). Springer-Verlag, New York.

Kelly, D.H. (1961). Visual responses to time-dependent stimuli. I. Amplitude sensitivity measurements. J. Opt. Soc. Amer. 51, 422-429.

Kelly, D.H. (1971). Theory of flicker and transient responses. I. Uniform fields. J. Opt. Soc. Amer. 61, 537-546.

Lamb, T.D. and Simon, E.J. (1976). The relation between intercellular coupling and electrical noise in turtle photoreceptors. J. Physiol. 263, 257-286.

Lamb, T.D. and Simon, E.J. (1977). Analysis of electrical noise in turtle cones. J. Physiol. 272, 435-468.

Shapley, R. and Enroth-Cugell, C. (1984). Visual adaptation and retinal gain controls. In Progress in Retinal Research 2. (eds. N.N. Osborne and G.J. Chader). Pergamon Press, Oxford.

Simon, E.J., Lamb, T.D. and Hodgkin, A.L. (1975). Spontaneous voltage fluctuations in retinal cones and bipolar cells. Nature 256, 661-662.

A New Concept for the Contribution of Retinal Colour-opponent Ganglion Cells to Hue Discrimination and Colour Constancy: The Zero Signal Dectector

E. Zrenner

Max-Planck-Institute for Physiological and Clinical Research,
Parkstrasse 1, D-6350 Bad Nauheim, German Federal Republic

INTRODUCTION

It is still not well understood which neuronal mechanisms are responsible for the large variety of colours we can discern in our visual space under various conditions of illumination. The recordings from a few hundred retinal neurons do not permit to explain or even to speculate reasonably on the neuronal basis of colour perceptions; the neurophysiological link is still missing, although recent investigations give reason to hope (Livingstone and Hubel, 1984). However, it is certainly possible to go the reverse way, that is to look out within perceptive phenomena for the action of neuronal mechanisms observed in individual retinal cells, which beyond any doubt contribute to building up these perceptions.

This paper will discuss to what extent retinal mechanisms can serve to enhance colour discrimination, i.e., contribute to the typical shape of the hue discrimination function. It shall also be considered to what extent temporal characteristics of retinal neurons determine colour vision properties and how retinal circuits can contribute to colour constancy. A new concept based on a cortical zero signal detector will be presented and its implications for colour perception will be discussed.

METHODS

The following considerations are based upon data on retinal ganglion cells of Rhesus monkey, 365 of which I recorded in a longstanding and fruitful collaboration with Professor Peter Gouras in Bethesda (Gouras and Zrenner, 1979, 1981; Zrenner and Gouras, 1981, 1983) and 265 of which I recorded in my laboratory in Bad Nauheim, partly together with Dr. Marion Wienrich (Wienrich and Zrenner, 1984). The methods are described extensively in my monograph (Zrenner, 1983a). In short, a trepanation of the sclera was performed in anesthetized and paralyzed Rhesus monkey. A metal cannula was introduced, through which a glass micropipette was advanced by a microhydraulic drive, observed through a conventional fundus camera. Monochromatic test- and adapting lights were projected onto the retina through this fundus camera in Maxwellian view. The responses were amplified and stored on magnetic tape for subsequent evaluation.

The prerequisites: Individual cone mechanisms, cone opponency and colour

If selective chromatic adaptation is used, the spectral sensitivity functions of individual cone mechanisms that have input to a ganglion cell, can be isolated. Such action spectra, based on a threshold criterion (increment or decrement of frequency of action potentials) fit reasonably well the pigment absorption spectra of the three spectrally different cones of Rhesus monkey retina (see Wienrich and Zrenner, 1984), called red (R), green (G) and blue (B) sensitive cones in the following.

In presence of white adapting light, spectrally different cones can interact and thereby produce the well known antagonistic response characteristics of colour-opponent ganglion cells illustrated in Fig. 1. A ganglion cell's (Gc) circuitry is shown on the left, the receptive field (RF) in the centre, and the responses on the

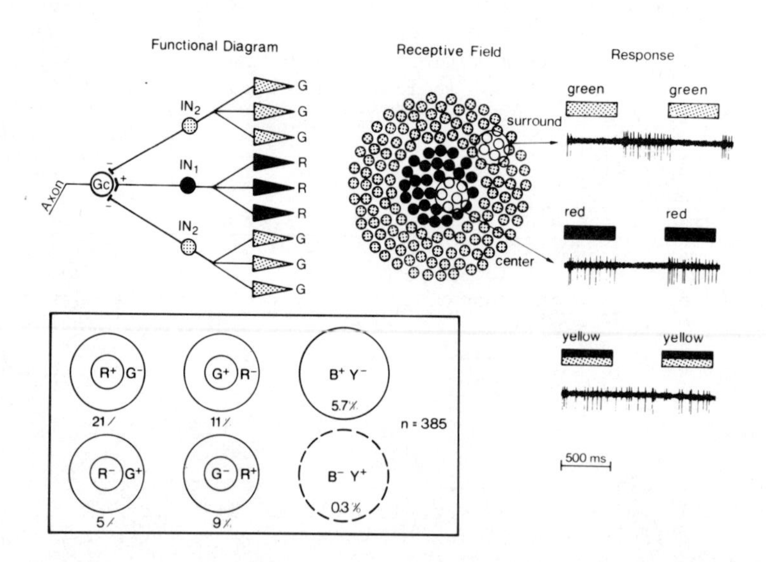

Fig. 1: Simplified functional diagram, receptive field structure and responses of a colour-opponent ganglion cell. Schematical structure and occurrence (in percent) of the six most common varieties of colour-opponent ganglion cells is shown in the inset figure (modified from Zrenner, 1983a).

right. Red light (black bar) is most effective for R-cones that excite (+) the cell through interneurons (IN_1) in the centre of the RF, so that the cell responds with a tonic discharge. Green light (dotted bar) is most effective for the G-cones in the surround; they inhibit (-) the cell through other interneurons, probably via horizontal and bipolar cells. Yellow light (black/dotted bar) has an about equal strength for the receptors that build up centre and surround, so that excitation and inhibition cancel each other.

Could a zero signal detector code for colour?

Imagine a scientist sitting in a room next to the laboratory, who has an oscilloscope and watches the records of the red/green-opponent retinal ganglion cell shown on the right of Fig. 1. Let's assume he does not know anything about the qualities of the light stimulus, but is expected to guess the colour or spectral locus of the stimulus. This comes close to the task a cortical cell might have, looking at the action potentials of a retinal ganglion cell's axon. It is of course impossible for the scientist to make a proper guess, since he does not know whether the series of action potentials he sees on the oscilloscope constitutes an ON-response or an OFF-response nor does he know whether there was a light stimulus occurring at all since "no-light" or a strong stimulus near the neutral point (e.g. yellow, Fig. 1, lowermost record) produce identical "zero"-responses. However, if there is one more piece of information, probably from a second electrode that records the activity of another, spectrally different ganglion cell, our hypothetical scientist (or the cortical cell, respectively) could do already a much better job. If the cell with the second electrode would tell him that there was a change of illumination, while the first electrode records a discharge in the red/green-opponent cell, he still would not know whether a red or a green light was falling upon the cell's receptive field since he would not know whether the train of action potentials he sees is a decrement following a green light or an increment during red illumination; both conditions produce the same response. However, one thing he could detect almost perfectly: If the second electrode tells him "there was an event", while the first electrode in the red/green-opponent cell indicates "no response" he could say for sure that the full field stimulus must have been composed of the neutral point wavelength "yellow". He knows that the response is zero, if red and green cones are stimulated about equally, be it a monochromatic light or any appropriate metameric mixture. Our hypothetical scientist would then act like a zero signal detector: he gives a signal when he makes a zero reading which in the above case indicates that there is a yellow light stimulus.

To estimate his performance we would have to look at the occurrence and spectral sensitivity of the cells in question. 66% of our population are tonic ganglion cells with a sustained response, the distribution of which is shown in the inset if Fig. 1. 46% of the cells are red/green-antagonistic (2/3 ON-centre, 1/3 OFF-centre), while 6% of the cells fall into the blue/yellow antagonistic category (see details in Zrenner, 1983a). Additionally, 14% of the cells belong to the broadband non-antagonistic cells.

The action spectra of the three main groups are shown in Fig.2. The reciprocal of the threshold irradiance (Quanta $\cdot$ $s^{-1} \cdot$ μm^{-2}) is plotted against wavelength; the constant threshold response is either a small impulse increment for the ON-response (open symbols, excitation) or a decrement for the OFF-response (closed symbols, inhibition). The typical red/green-antagonistic cells in the foveolar region have a sensitivity maximum for excitation near 610 nm and for inhibition near 500 nm (or vice versa) with a neutral point near 560 nm. The cells' bandwidth is smaller than that of a pigment absorption spectrum; it can, however, rather nicely be described by linear subtraction of pigment absorption spectra (/R-G/), as indicated by a dashed line in Fig. 2 (top). The second cell type (blue/yellow-antagonistic) is almost exclusively excited by short-wavelength light, with a sensitivity maximum near 440 nm for the excitatory response and 560 nm for the inhibitory one. The action spectra can be described by a subtraction: /B-(R+G)/. Both cell types have a "weakness": their responses are qualitatively different

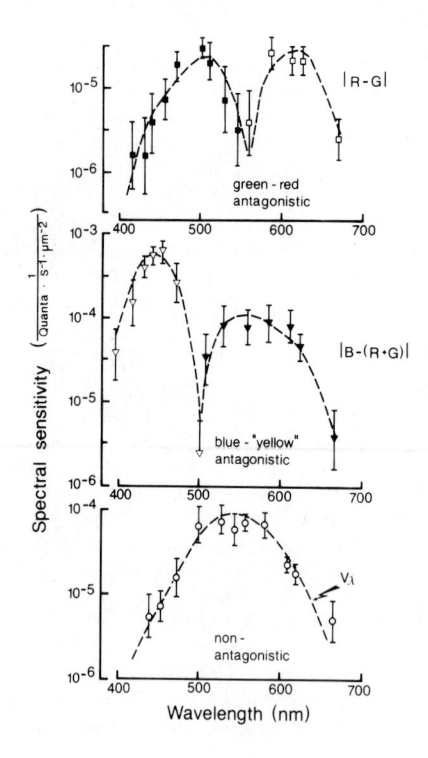

Fig. 2: Action spectra of 8 red/green-antagonistic and 6 blue/yellow-antagonistic opponent tonic ganglion cells as well as of 3 non-antagonistic broadband tonic cells, recorded in presence of white Ganzfeld background (20,000 td); mean ± S.D.. Open symbols indicate thresholds of excitatory responses, i.e., just detectable spike frequency increments in response to increments of the test flashes' (15° in diameter) irradiance. Closed symbols indicate thresholds of inhibitory responses, i.e., just detectable spike frequency decrements in response to increments of irradiance. Broken lines (top and centre) indicate a linear subtraction of pigment absorption spectra of the cone types indicated on the right of each graph (from Zrenner, 1983b).

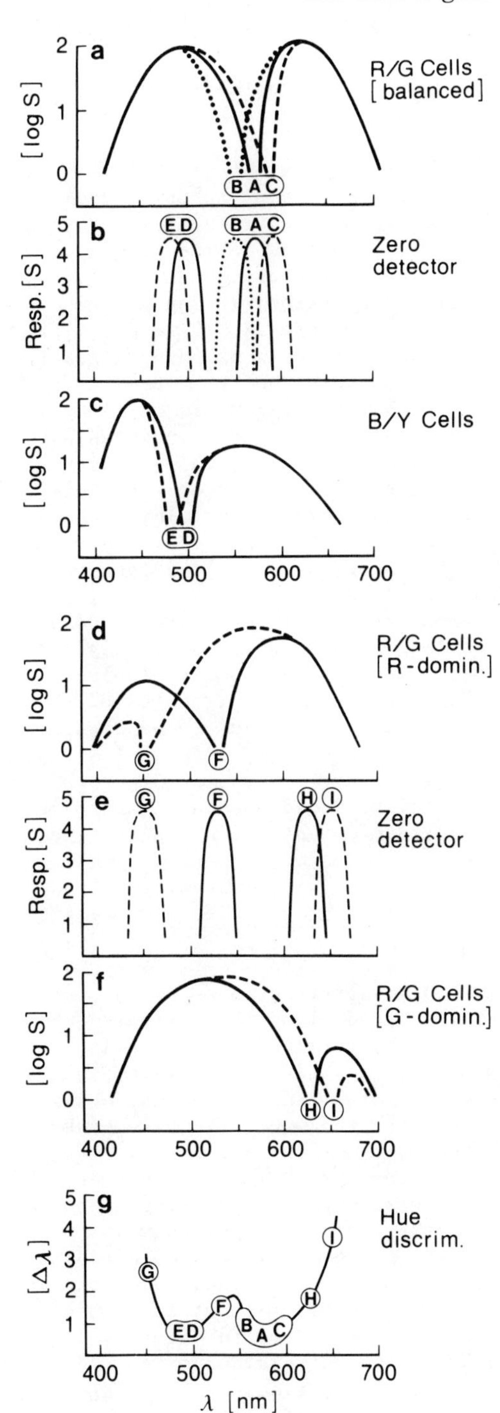

Fig. 3: Action spectra of colour-opponent cells with balanced input (a,c) from spectrally different cones, together with the sharply tuned spectral sensitivity of the proposed cortical zero detectors (b). The zero detector fires strongest, when a light stimulus is detected, to which retinal colour-opponent cells do not respond. This is the case, where the neutral point condition is reached, i.e., when the spectrally different opponent cone mechanisms are stimulated equally. Since the input of these cones can be of different weight, neutral points can vary in their spectral position (A-E), generating zero detectors (b) with a sensitivity maximum in that spectral region (A-E). The physiological variations of the strength of cone input from strongly R-cone dominated (d) to strongly G-cone dominated cells (f), as lined out in Fig. 4, produce neutral points (F-I) at all other spectral regions, thereby building up zero detectors (F-I) in these regions (e). The hue discrimination then depends on the smallest difference in spectral loci of two neighbouring zero detectors, consequently being best in the spectral regions of highest incidence of these detectors (i.e., 490 nm and 575 nm).

which is inconvenient for a comparison of the brightness of two spectrally different stimuli. The third type of tonic non-antagonistic cells can compensate for this "weakness". The action spectra of these cells follow the V_λ-function, reflecting a rather additive processing of spectrally different cone mechanisms.

How would our hypothetical scientist perform, if the broadband cells would tell him that there was an event, while he simultaneously looked at the above described opponent cell's output? The majority of red/green-opponent cells have neutral points near 575 nm, forming a cluster with only slight neutral point deviations as schematically shown in the three action spectra (A,B,C) of Fig. 3a. The scientist's zero-readings on these cell types would give him rather precise information about stimuli with spectral loci near 575 nm. His hue discrimination function (Fig. 3g) would be rather good in this region since the dense cluster of neutral points (A,B, C) ensures that slight changes of few nanometers in the stimulus wavelength would immediately cause a zero reading in another cell (Fig. 3b, A,B,C), the neutral point of which is coinciding with the new wavelength. If the scientist would look at the blue/yellow-opponent cells' output (Fig. 3c), he accordingly could detect very precisely blue/green lights with a spectral locus near 500 nm, just by signalling zero readings via the zero signal detectors E and D (Fig. 3b). His hue discrimination (Fig. 3g) should be slightly worse in this region (E,D) than near 575 nm, since the number of blue/yellow-opponent cells is small. Consequently, the distances of the spectral loci of two stimuli must be larger in order to be signalled by two different cells.

What about the other wavelengths? He would need neutral points at all wavelengths; indeed, to a certain extent there are such cells in the primate retina.

Variations in colour-opponency, neutral point position and zero detector tuning

Fig. 4a shows the spectral sensitivity functions of six such colour-opponent cells (arbitrarily shifted along the ordinate), all belonging to the red/green variety, as determined by chromatic adaptation, which reveals the spectral sensitivity of the cones that are connected to this cell. There is a wide range of variations in spectra and neutral points. In presence of a neutral, white background of 30.000 td the cell in the centre shows balanced inputs; a maximum of its excitatory response (open symbols) occurs near 600 nm, a maximum of its inhibitory response near 500 nm, with a neutral point in between, where the cells' threshold is high. Only 14% of our tonic cells showed such balanced opponency, where inhibition and excitation covered about half the spectrum and had similar thresholds (Zrenner and Gouras, 1983). The majority of the tonic cells (45%) were dominated either by the red (top in Fig. 4a) or the green sensitive cone mechanism (bottom of Fig. 4a) with the neutral point shifting accordingly. Thus one finds a wide range of

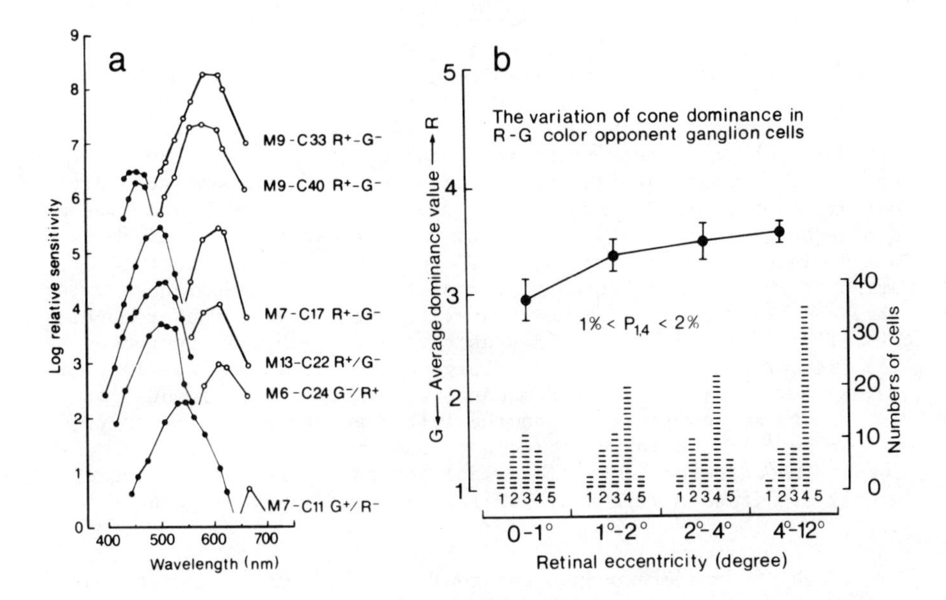

Fig. 4: a) Action spectra of 6 red/green-opponent cells, recorded under identical conditions (white background, 20.000 td, Ganzfeld stimuli of 200 ms duration). The action of R-cones is indicated by open symbols, that of G-cones by closed symbols.

b) The distribution of cone dominance (1-5, i.e., from extremely green to extremely red cone-dominated cells) at different retinal eccentricities; standard error of the mean (SEM) indicated by vertical bars (modified from Zrenner, 1983a and Zrenner and Gouras, 1983).

variation in the strength of the red and green sensitive cones, contributing to a colour-opponent cell. Neutral points can be found almost in any part of the visible spectrum (between 420 and 650 nm) and sensitivity maxima of excitation and inhibition can differ almost by two log units. Consequently, there would be sufficient variations in neutral point wavelengths to build up zero detectors throughout almost the entire visible spectrum.

To quantify this phenomenon, we introduced a scale of cone dominance (Fig. 4b), ranging from 1 to 5, where a dominance value (D) of 3 indicates that the strength of both the opposing mechanisms is about balanced. D = 4 indicates dominance of red sensitive cones, with colour-opponency still seen on white background, being extremely dominant at D = 5. Correspondingly D = 2 indicates dominance of the green sensitive cone mechanism, being extremely dominant at D = 1. When we examine the incidence of each of these five groups in terms of numbers of cells at different retinal eccentricities (right hand ordinate in Fig. 4b), an interesting picture of the distribution of cone opponency emerges (Zrenner and Gouras, 1983): Even in the foveolar region all varieties of dominance (1 to 5) are present, but the largest fraction of cells has dominance values of 3 (balanced opponency). Notice that in all areas of the central retina representatives of each dominance values are seen. However, as one moves away from the fovea, the proportion of cells dominated by the red sensitive cone mechanism increases considerably, as indicated by a large number of cells with a dominance value of 4. If the average dominance value of each individual area is calculated (as indicated on the left hand ordinate), it shifts gradually from 2.8 in the foveola to 3.4 in the periphery. The difference between the first (foveolar) and the fourth group becomes highly significant with a probability of error only between 1% and 2%.

Thus the red/green-opponent tonic ganglion cells seem to form a large heterogeneous group, with the entire range of heterogeneity represented in each retinal area and with cone dominance changing regularly with retinal eccentricity; R-cones become increasingly dominant towards retinal periphery; the consequences of this shift are discussed by Zrenner (1983a). Cells receiving input from blue sensitive cones are quite different; the spectral sensitivity functions of all the cells measured so far (Zrenner and Gouras, 1981; Wienrich and Zrenner, 1984) are strikingly similar in shape and absolute sensitivity.

What role can an arrangement of varying opponency between cone mechanisms play in the visual system? If an equiluminant red/orange border is moved over the receptive field of the cells shown in Fig. 4a, not much of a signal would be produced since red and orange excite these cells about equally, except for the lowermost cell shown in Fig. 4a. This latter cell would respond to such a moving border very strongly with a polarity reversal, since its neutral point lies in between the two spectral loci that form the chromatic border. Consequently, if an object is large enough or moved over the retina, there is always a small population of cells, the neutral point of which lies in between the spectral loci of the two hues forming a chromatic border. These cells and only these cells would respond maximally, changing from excitation to inhibition or vice versa, thus signalling optimally a chromatic border moved through the cells' receptive field. Therefore the inhomogeneity of the red/green-opponent tonic system can enhance greatly the capability of the visual system to respond maximally to a large variety of chromatic borders within each area of visual space, but particularly in the fovea, where cone dominance values are most evenly distributed.

As lined out above, spectrally there are two clusters of cells in the fovea with neutral points most densely packed near 575 and 500 nm. This is indicated by the neutral points and zero detectors A, B and C in Fig. 3a and b, as well as by neutral points and zero detectors in Fig. 3c and b. The frequency of cells with neutral points in other spectral regions is considerably less, such as the neutral points G and F in Fig. 3d that are formed by R-cone dominated colour-opponent cells or neutral points H and I in Fig. 3f, formed by G-cone dominated colour-opponent cells. Consequently the number of zero detectors of the types shown in Fig. 3e would be smaller than that shown in Fig. 3b. The hue discrimination would be of course best at spectral loci where the neutral point cluster is most dense. Hue discrimination would depend on the number of zero detectors in a fixed spectral range, since the distance of the spectral loci of two objects must be wider in order to be detected by two different zero detectors the spectral distance of which is wide. As a consequence, the shape of the hue discrimination function (Fig. 3g) is determined by the density of zero detectors in the spectrum: Hue discrimination is very good in spectral regions that are covered by the most common neutral points A-D in Fig. 3g (and the corresponding zero detectors), while it is worse at spectral regions covered by the least common neutral points and zero detectors, respectively (F-I in Fig. 3g)

This model of wavelength discrimination is based on the variation of neutral points in colour-opponent retinal ganglion cells. It would require cortical cells which serve as zero detector, signalling when the neutral point condition is achieved. These cells - if they exist - would be marked by very narrow spectral sensitivity; they would be most common in the fovea with higher cluster density near 500 and 575 nm, shifting towards longer wavelengths for eccentric stimulation. Due to the physiological variations in neutral point position, sharply tuned (spectrally asymmetric) zero detectors could occur in any part of the visible spectrum, except at the very short and very long wavelengths; such cells were described e.g. by Zeki (1980). Since we can discern only about 160 hues in the visible spectrum, the number of such cells could indeed be limited. Their distribution in the visible spectrum could explain the particular shape of the wavelength discrimination function

How to build a cortical zero detector

As shown in Fig. 5, such a cortical zero detector can easily be built up: A cortical cell that has an excitatory input (+) from a broad band tonic cell and an inhibitory one (-) from a colour-opponent cell, could fire only under the following conditions: Firstly, the excitatory input has to be activated (i.e., only if a light stimulus is present), and secondly, the opponent cell's input has to be in a silent neutral point condition. Since responses at all other wavelengths cancel each other, such a zero detector would be very sharply tuned. Fig. 5 serves as a schematic example; such a circuit

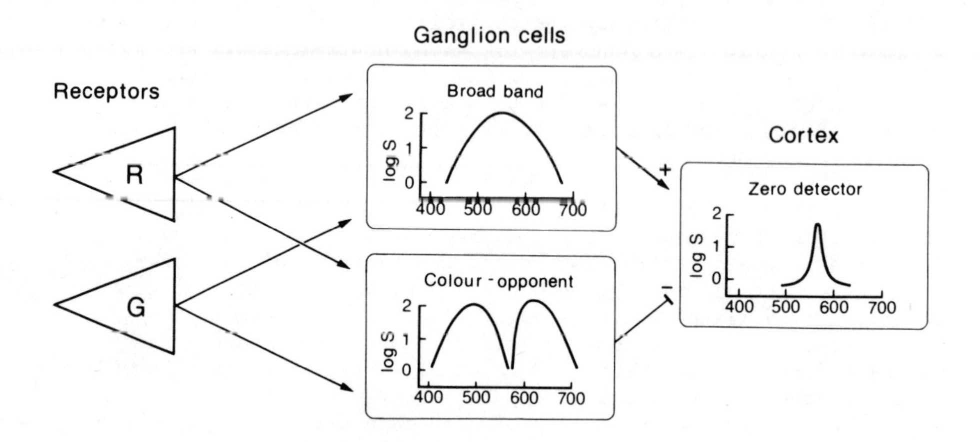

Fig. 5: Tentative model of the wiring of the proposed cortical zero detectors. Red sensitive cones (R) and green sensitive cones (G) form either broad-band ganglion cells (top) or colour-opponent cells (bottom), the spectral sensitivity of which is sketched in the box. The colour-opponent ganglion cell's inhibition (-) on the broad-band cell's excitatory input (+) to a zero detector is reduced only with stimuli that match neutral point conditions (575 nm in this example). This is a simplified scheme. Since the opponent cell produces a train of pulses in one part of its spectrum during the presence of light, in the other part after the offset of light, actually three types of retinal cells could be involved: Either one colour-opponent cell and a pair of broad-band cells with differently weighted cone input or one broad-band cell and a pair of colour-opponent cells with reversed signs in their two spectral regions. Further explanations in the text.

would work as well with R- or G-cone dominated colour-opponent ganglion cells (Fig. 4) or with blue/yellow-opponent cells. In any case it produces a series of zero response detectors, that are tuned to the retinal colour-opponent cells' neutral points.

The timing of ON- and OFF-responses of cells is of course critical in such a circuit. Due to the axonal characteristics of ganglion cells, graded responses can have no effect; therefore the appropriate retino-cortical signal can only consist of a train of action potentials, irrespective of the fact whether it was elicited by the onset or by the offset of light. Since one of the wavebands of the colour-opponent cell in Fig. 5 corresponds to responses at the onset of a stimulus, the other waveband to responses at the offset of the stimulus, at least three retinal cells need to be involved in generating the appropriate retinocortical signal for the zero detector. There are several alternatives for the realization of the circuit lined out in Fig. 5: The colour-opponent signal, the spectral sensitivity of which is shown in the lower half of Fig. 5 would have to involve a pair of retinal colour-opponent cells with reversed polarity such as a R-OFF/G-ON and a R-ON/G-OFF pair, the excitatory responses of which are inhibiting the response of one or more broad-band cells (shown on top of Fig. 5) that have input to the zero detector. Such a circuit would not easily by applicable to the B/Y-opponent cells, since B^+/Y^- and B^-/Y^+ cells do not occur in equal proportions (Zrenner and Gouras, 1981).

Another possibility would be to have only one type of balanced colour-opponent cell connected with a pair of cells with broader spectral sensitivity of different polarity; these broad-band cells could still show some opponency at the end of the spectrum as observed in the R- and G-cone dominated colour-opponent cells shown on top and on bottom of Fig. 4. For example, consider the cell type shown in Fig. 4: The excitatory responses generated at light onset by the ON-band of the colour-opponent cell (M7-C17, R^+/G^-) via the R-cone would be in balance with the ON-band of the broad band R-cone dominated cell (M9-C33, R^+/G^-); subtraction of the ON-part of both spectral sensitivity functions would result in a zero detector tuned near 550 nm. Similarly the excitatory responses generated after the light offset by the OFF-band of the colour-opponent cell (M7-C17) via the G-cone would be in balance with the simultaneously occurring OFF-responses of the broad-band G-cone dominated cell (M6-C24, G^-/R^+, or if polarity reversal were already included in the cell response, with M7-C11, G^+/R^-); the subtraction of the OFF-part of both spectral sensitivity functions would again result in a sharply tuned zero detector for OFF-responses at 550 nm.

A third, rather elegant way would be to simply balance an R-cone dominated (e.g. M9-C33, R^+/G^-) opponent cell and a G-cone dominated one (M7-C11, G^+/R^-) against each other; the difference spectra in such a circuit, however, would result in a zero detector that has two sharply tuned bands, one for ON-responses in the longwave region and one for OFF-responses in shorter wavelength regions.

Whichever version of the circuit is utilized, it would result in a zero detector that is organized in a double-opponent way, as described in many cells in the visual system, most recently by Livingstone and Hubel (1984). It is also self-evident that a general circuit of the kind shown in Fig. 5 would work with blue/yellow-opponent cells as well.

Spectral tuning of zero detectors would depend strongly on temporal and spatial conditions

Why do we lose good hue discrimination with small test fields or flickering pairs of stimuli? If the neutral zone of a colour-opponent cell is broadened or lost, e.g. when small stimuli are

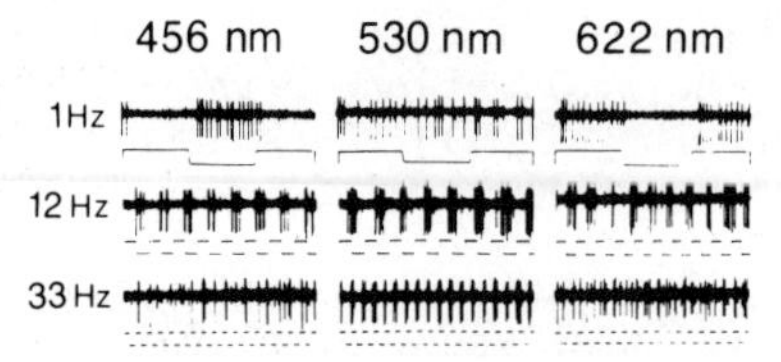

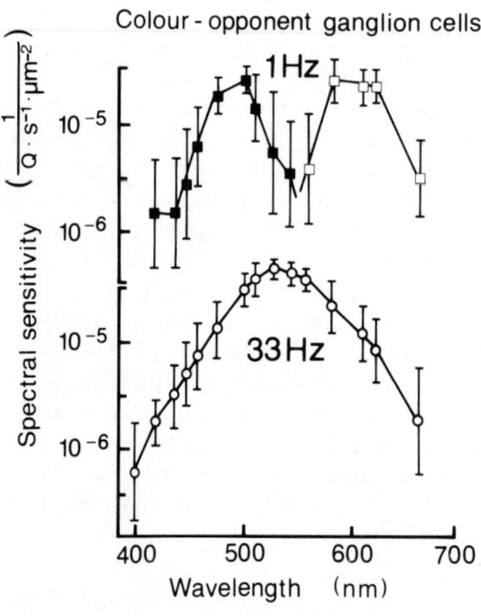

Fig. 6: Action spectra of nine red/green-opponent ganglion cells (mean values ± S.D.), based on a flicker threshold criterion at low (1 c/s, triangles) and at high (33 c/s, circles) frequencies of stimulation using a white adapting light of 30.000 td (Ganzfeld-stimulation). An example of the responses is shown on top. Only red/green-opponent cells with balanced colour-opponency were taken, i.e., cells with dominance values of 3 as described in Fig. 4. Curves were combined at their mean maximum sensitivity (from Gouras and Zrenner, 1979, as reported by Zrenner and Gouras, 1978).

used (Wiesel and Hubel, 1966) or flickering stimuli are applied, the spectral tuning of a zero detector might become worse or eliminated. An example of the influence of flicker rate on the neutral point is shown in Fig. 6 (from Gouras and Zrenner, 1979), which (on top) illustrates the responses of a colour-opponent ganglion cell with a red-sensitive ON-centre (R+) and a green-sensitive OFF-surround (G-) at three different wavelengths and three different flicker rates (1, 12 and 33 c/s). At low frequencies, when the light stimulus goes on, the cell is excited by long wavelengths (622 nm) and inhibited by short wavelengths (456 nm). When the light goes off, inhibition occurs at 622 nm while excitation is observed at 456 nm. Centre and surround responses are almost 180 degrees out of phase. Light from the middle of the spectrum (530 nm) is practically ineffective because it stimulates the opposing cone mechanisms about equally. At medium flicker rates the effectiveness of 530 nm stimuli begins to increase; at high flicker rates this cell becomes most effective at the former neutral point wavelength, while responses to short- and long-wavelength stimuli are almost fused. This phenomenon occurred in all colour-opponent cells that were studied.

The corresponding spectral sensitivity functions are shown in the lower half of Fig. 6, based on flicker threshold criteria in nine red/green colour-opponent ganglion cells at low and high frequencies (1 Hz and 33 Hz). At low frequencies of stimulation, all cells show two sensitivity maxima, with a neutral zone in between: The cortical zero detector lined out above would be well tuned under such a condition, and indeed colour discrimination is good at low frequency flicker. At high flicker frequencies (33 Hz), the wavelengths of the neutral zone become most effective in all cells, so that colour-opponency is lost. The spectral sensitivity function becomes broad-band and single-peaked and resembles precisely that of a luminance detecting cell. The mechanism of this change in action spectra is based on a phase shift between centre and surround of colour-opponent cells (details see Zrenner and Gouras, 1978; Gouras and Zrenner, 1979; Zrenner, 1983a). A similar mechanism applies to blue/yellow-opponent cells.

Consequently the capability of a cortical narrow-band zero signal detector to code for a small spectral region would strongly depend on the temporal properties of retinal colour-opponent ganglion cells. Due to the lack of a neutral point in the retinal colour-opponent cell at higher frequencies of stimulation, the cortical zero detector could not work anymore under such conditions, since it cannot be supplied with its appropriate input, i.e., a neutral point zero signal from retinal ganglion cells. Indeed colour discrimination for flickering stimuli worsens or disappears. Consequently, the function of the zero detector would be frequency-dependent. A similar loss of hue discrimination would occur for small spatial stimuli (spots), which abolish the neutral point in colour-opponent retinal cells, stimulating centres only.

Zero signal detectors and colour constancy

This concept of zero detectors, the spectral sensitivity of which depends on the neutral point position of colour-opponent retinal cells can contribute also to mechanisms of colour constancy, as lined out in the following: The zero detector - by definition - produces a strong signal whenever the light conditions are such as to provide balanced signals from both the cone mechanisms that feed a colour-opponent cell, in other words, if the output of the cell reaches the neutral point condition. Consequently, the zero detectors just signal zero responses independently of the various light conditions that can produce such a zero response. In fact, the position of the neutral point depends very much on the spectral composition of the surrounding illumination, as shown in Fig. 7. If the background illumination is changed from white to a longer wavelength illuminant, the neutral point of a red/green-opponent ganglion cell can shift from 510 to 570 nm. Consequently, the cortical zero detector that is connected to this cell would change its sharp spectral tuning peak from 510 to 570 nm. Interestingly, it would be now well tuned for the new appearance of the object: The zero signal detector that before interpreted 510 nm as "his" specific spectral locus, e.g. blue/green, now interpretes 570 nm as his specific spectral locus, that is, blue/green too. This indeed would be necessary

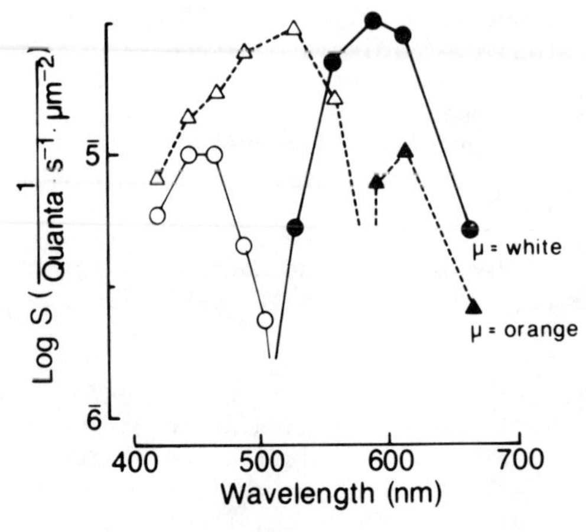

Fig. 7: Action spectra of an R-cone dominated red/green-opponent ganglion cell based on response threshold criteria in presence of two different illuminants, white (circles, 5500 k) and orange (triangles, Schott OG 5). Ganzfeld stimuli (30.000 td) of 200 ms duration. Excitatory responses are indicated by open symbols, inhibitory ones by closed ones. Note the shift in neutral point position into the spectral direction of the illuminant (modified from Zrenner, 1983a).

for colour constancy, since the spectral locus of the object in the colour space has been shifted by the long-wavelength illuminant also to longer wavelengths, according to Akita, Graham and Hsia (1964). Indeed, an area that appears blue/green in a Mondrian picture under a white illuminant would still appear blue/green under a long-wavelength illuminant, as demonstrated by Land (see this volume). Such a mechanism, that ensures identical appearance of an object, independent of the wavelength composition of the illuminant would provide colour constancy since - in Helmholtz' (1866; 1911) terms - it "discounts the illuminant colour". A blue/yellow-opponent cell would behave very similarly: A change of the illuminant from white to orange weakens the long-wave cone input, so that the blue sensitive cone's input becomes more sensitive (for an example see Zrenner, 1983a) and shifts the neutral point to longer wavelengths. Also in those cells the neutral point shifts into the direction of the background wavelength, exactly what it should do to give identical signals despite the new spectral composition. This retinal mechanism would work as well with bluish illuminants, e.g. in a red/green-opponent cell: The more strongly bleached green sensitive cone's input to a colour-opponent cell becomes less sensitive if a short-wavelength illuminant is present; the red-sensitive cone's input becomes more sensitive, since the opponent signal provided by G-cones is weaker; the neutral point again shifts in the direction of the illuminant in such cells (see examples in Zrenner, 1983a). This concept could be extended to include Helson's (1938), Judd's (1940) and Helson and Michel's (1948) experiments on colour appearances under different conditions of illumination. The necessary quantitative treatment, however, would go beyond the scope of this paper. Nevertheless, if this concept of a zero signal detector should be applicable, the zero signal's spectral locus provided by the retinal colour-opponent cells would move in the spectral direction of the illuminant and thus be properly adjusted to the new physical properties of an object that is illuminated by this illuminant.

In Summary:

The spectral properties of colour-opponent cells recorded from monkey retina change depending on retinal eccentricity (Zrenner and Gouras, 1983) and, in individual cells, depending on the temporal properties of stimulation (Zrenner and Gouras, 1978; Gouras and Zrenner, 1979) as well as on the spatial distribution of light stimuli and on the spectral properties of the illuminant (Wiesel and Hubel, 1966; see also Zrenner, 1983a); red/green-opponent cells provide a spectrally heterogeneous group with neutral points varying from 420 to 650 nm; red sensitive cones become more dominant in red/green-opponent ganglion cells with retinal eccentricity; colour-opponency is lost, when flicker rate is increased, and neutral points shift into the direction of the illuminant's spectral locus.

The concept of a cortical zero response detector proposed here provides spectrally tuned signals to higher order neurons. Such a detector should provide an output signal only if it detects the occurrence of light stimulation in any spectral region that, however, had no effect on the particular retinal colour-opponent cell which the cortical zero detector is connected to. Such a situation can occur only in the very narrow spectral region of the neutral point of a colour-opponent cell. The neutral point condition itself, however, can be achieved by physically very different lights, that produce a cancelling balance of the colour-opponent cell's inputs provided by spectrally different cones. All these physically very different lights (monochromatic or broad-band) would generate the same signal via the zero detector and therefore appear identical.

This proposal has the following advantages:

1. Spectral loci are coded by sharply tuned signals; thereby the existence of spectrally sharply tuned cortical cells would be explained.

2. The varying density of neutral point distribution of R/G- and B/Y-opponent cells in the visible spectrum correlates with the hue discrimination function.

3. It would explain why hue discrimination deteriorates with increasing flicker rate, decreasing stimulus duration and decreasing field size, since all these conditions eliminate or broaden the neutral point zone (see Zrenner, 1983a).

4. It could explain flicker-induced colours (as in the case of the Fechner-Benham top); i.e., phase shifts between centre and surround of colour-opponent cells provide an artificial zero response (see Zrenner, 1983a) that necessarily would be interpreted by the zero detector as hue.

5. With such an arrangement, retinal colour-opponent cells could also contribute to colour constancy, since their neutral point is shifted by chromatic illuminants in the direction of the illuminant's spectral locus. Thereby the illuminant would shift the chromatic tuning of the whole set of zero response detectors in that direction in which the spectral loci of the illuminated objects have been shifted; consequently, the zero signals would be invariant for identical objects, irrespective of the illuminant, at least within a large range.

6. It could explain the Bezold-Brücke phenomenon; neutral points of colour-opponent mechanisms become more sharply tuned at 500 and 575 nm, when the background illumination is increased (Sperling and Harwerth, 1971). Consequently, at high levels of illumination, the main cluster of zero detectors is active in two regions: blue/green and yellow.

This attempt is meant to tentatively resolve problems that arise from experimental facts in retinal ganglion cells, with the question in mind "why do we have colour-opponency in the retina?" It is this type of question that the meeting was designed to address, as I understood Professor Ottoson.

Acknowledgements

I am very grateful to Prof. P. Gouras, Dr. A. Valberg, B. Olsen and Dr. Th. Schneider for critical discussions on the concept presented here as well as to Mrs. H. Breitenfelder for carefully typing the manuscript and to Mrs. H. Schneider for preparing the drawings and photographs.

References

Akita, M., Graham, C.H. and Hsia, Y. (1964). Maintaining an absolute hue in the presence of different background colors. Vision Res. 4, 539-556.

Gouras, P. and Zrenner, E. (1979). Enhancement of luminance flicker by color-opponent mechanisms. Science 205, 587-589.

Gouras, P. and Zrenner, E. (1981). Color vision: A review from a neurophysiological perspective. In Progress in sensory physiology vol. I. (eds. H. Autrum, D. Ottoson, E.R. Pearl and R.F. Schmidt). Springer, Heidelberg, pp. 139-179.

Helmholtz von, H.L.F. (1866). Handbuch der physiologischen Optik, vol. II. Voss, Leipzig.

Helmholtz von, H.L.F. (1911). Handbuch der physiologischen Optik, vol. II. Voss, Leipzig.

Helson, H. (1938) Fundamental problems in color vision. I. The principle governing changes in hue, saturation and lightness of non-selective samples in chromatic illumination. J. exper. Psychol. 23, 439-476.

Helson, H. and Michels, W.C. (1948). The effect of chromatic adaptation and achromaticity. J. opt. Soc. Amer. 38, 1025-1032.

Judd, D.B. (1940). Hue, saturation, and lightness of surface colors with chromatic illumination. J. opt. Soc. Amer. 30, 2-32.

Livingstone, M.S. and Hubel, D.H. (1984). Anatomy and physiology of a color system in the primate visual cortex. J. Neurosc. 4, 309-356

Sperling, H.G. and Harwerth, R.S. (1971). Red-green cone interactions in the increment-threshold spectral sensitivity of primates. Science 172, 180-184.

Wienrich, M. and Zrenner, E. (1984). Cone mechanisms and their colour-opponent interaction in monkey and cat. Ophthalmic Res. 16, 40-47.

Wiesel, T.N. and Hubel, D.H. (1966). Spatial and chromatic interactions in the lateral geniculate body of the rhesus monkey. J. Neurophysiol. 29, 1115-1156.

Zeki, S. (1980). The representation of colours in the cerebral cortex. Nature 284, 412-418.

Zrenner, E. (1983a). Neurophysiological aspects of color vision in primates. Springer, Berlin.

Zrenner, E. (1983b). Neurophysiological aspects of colour vision mechanisms in the primate retina. In Colour vision: Physiology and psychophysics. (eds. J.D. Mollon and L.T. Sharpe). Academic Press, London, pp. 195-210.

Zrenner, E. and Gouras, P. (1978). Retinal ganglion cells lose color opponency at high flicker rates. Invest. Ophthalmol. Vis. Sci. (ARVO-Suppl.) 17, 130.

Zrenner, E. and Gouras, P. (1981). Characteristics of the blue sensitive cone mechanism in primate retinal ganglion cells. Vision Res. 21, 1605-1609.

Zrenner, E. and Gouras, P. (1983). Cone opponency in tonic ganglion cells and its variation with eccentricity in rhesus monkey retina. In Colour vision: Physiology and psychophysics. (eds. J.D. Mollon and L.T. Sharpe). Academic Press, London, pp. 211-223.

Colour Coding in the Primate Retinogeniculate System

P. Gouras

Department of Ophthalmology, Columbia University,
630 West 168 Street, New York 10032, USA

This presentation considers how information about color vision is coded in the retinogeniculate pathway in primates and how this information may be decoded by visual cortex.

At the start, it is useful to consider the three cone photoreceptor systems responsible for trivariant color vision in primates. Figure 1 highlights several important points. If the absorption spectra of the cone pigments are plotted on a fourth root of a wavelength scale (Dartnall et al, 1983), two cone types are found relatively close to one another near the longer wavelength end of the visible spectrum and a third at the shorter wave end. Human psychophysics shows that only the longer wavelength cone system is involved in high spatial resolution but both are involved in trivariant color vision and coarse spatial resolution. The mosaic of short wave (S) cones is coarser than that of the longer wave (M and L) cones, undoubtedly due to the problem of chromatic aberration.

A second point is that using a short wave cone in combination with only one type of long wave cone for color vision must be more useful for survival than using only two longer wave cone mechanisms. An hypothesis for why this is so has been given elsewhere (Gouras and Eggers, 1983; Gouras, 1984). Empirically we know that other mammals such as the cat, rabbit, ground squirrel as well as about 3% of human males rely mainly on a divariant form of color vision mediated by the S and one variety of longer wave cones. Absence of the S cone mechanism is very rare in man and probably most other mammals with color vision. The S and either the M or L cone mechanism alone provides an animal with hue (blue/yellow), saturation (white/black) and brightness contrasts. Adding a second long wave cone system provides only an additional hue (red/green) contrast. The S and one long wave cone provide color contrast across the entire visible spectrum, the white range; the M and L cones provide color contrast mainly in a narrower range, the yellow range. It is also important to realize that the contrast formed by

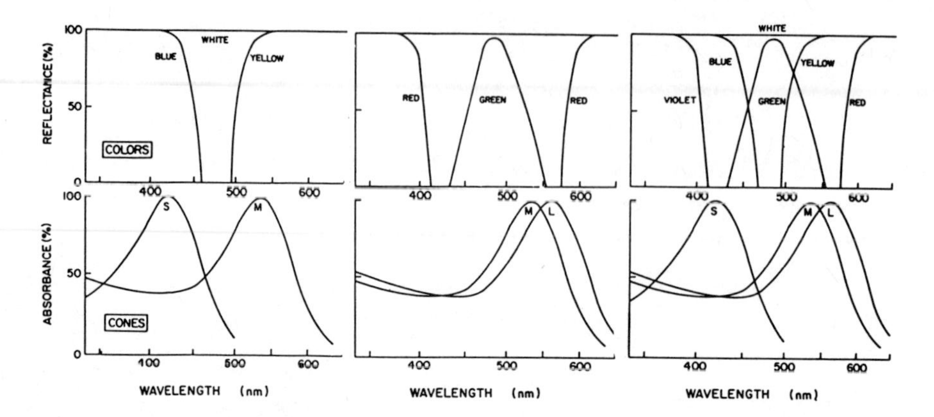

Fig. 1 Absorption spectra (below) of human cones plotted on a fourth root of wavelength scale (Dartnall et al, 1983) and the chromatic sensations (above) considered to be induced when each cone type is preferentially stimulated. The left set shows a divariant form of color vision in which the visible spectrum is split into blue, white and yellow regions. This system cannot easily distinguish greys from greens because greens, absorbing from both ends of the spectrum, can be simultaneously metameric to both S and M cones. The middle set shows a second system of divariant color vision which can offset this vulnerability for spectral contrast of the previous system. Here two longer wavelength sensitive cones (M and L) straddle the yellow region of the spectrum. The right set shows both sets combined as a trivariant form of color vision (Gouras and Eggers, 1983; Gouras, 1984).

the S and longer wave cones and that formed by the M and L cones appear to occur relatively independently of one another. This independence and the functional asymmetries of these mechanisms are relevant to the neural coding of color.

A major insight into color vision came from the introduction of single unit recording techniques to the visual system by Adrian in Cambridge and then Granit in Stockholm. This led to the discovery of color opponent interactions in single neurons of the fish retina by Gunnar Svaetichin here in Stockholm in the Fifties (Svaetichin, 1956). Color opponency is essential in the transformation of information from the cones to the central nervous system. Color opponency occurs when a neuron is excited by one cone mechanism and inhibited by a spectrally different mechanism. Such cells exist in the retina (Hubel and Wiesel, 1960) and lateral geniculate nucleus (DeValois, 1960) of primates. Color opponent cells are strongly excited by successive (temporal) color contrast. Turning off a color that preferentially affects the inhibitory cone input while simultaneously turning on a color that preferentially

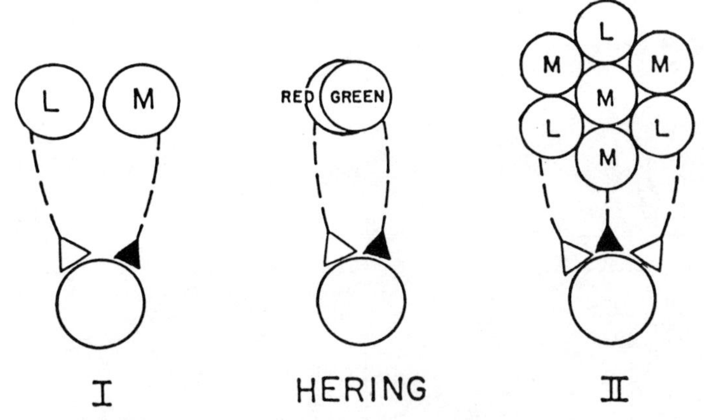

Fig. 2 Three forms of color opponent units. The form on the left occurs in vivo because spectrally different photoreceptors (M and L cones) are in different retinal regions. One cone sends excitatory (open triangles) and the other cone inhibitory (closed triangles) to the same neuron. The form in the middle represents the hypothetical red/green opponent channel of Hering in which redness opposes greenness in the <u>same</u> area of visual space. The form on the right shows how the retina (or brain) can approximate a Hering unit by having a postreceptoral neuron integrate the excitatory signals from one type (all M) and inhibitory signals from the other type (all L) of cone.

affects the excitatory cone input produces maximal excitation of such cells because an excitatory off-response is added to an excitatory on-response.

The discovery of these cells led to a revival of the Hering Theory of color vision. Most, if not all, color opponent primate retinogeniculate cells, however, are not equivalent to the hypothetical color opponent channels of Hering. A color opponent cell receives excitation from one point in space, usually a cone, and inhibition from another point in space, a different cone. Two spatially (and spectrally) distinct cones are needed, as a minimum, to form a single color opponent cell (Fig. 2, left). The Hering channels require color opponency in the <u>same</u> area of visual space (Fig. 2, middle). This is an important difference. The brain could mimic a Hering unit by forming a neuron that integrated signals from an intermingling group of different cones and thereby quasi-completely cover the same area of visual space with two different sets of cones (Fig. 2, right). This, however, would sacrifice the spatial resolution of the receptor mosaic. I believe the retinogeniculate system preserves this spatial resolution using the same cells that transmit L/M color contrast information. By doing this it conserves space in an otherwise crowded central nervous system.

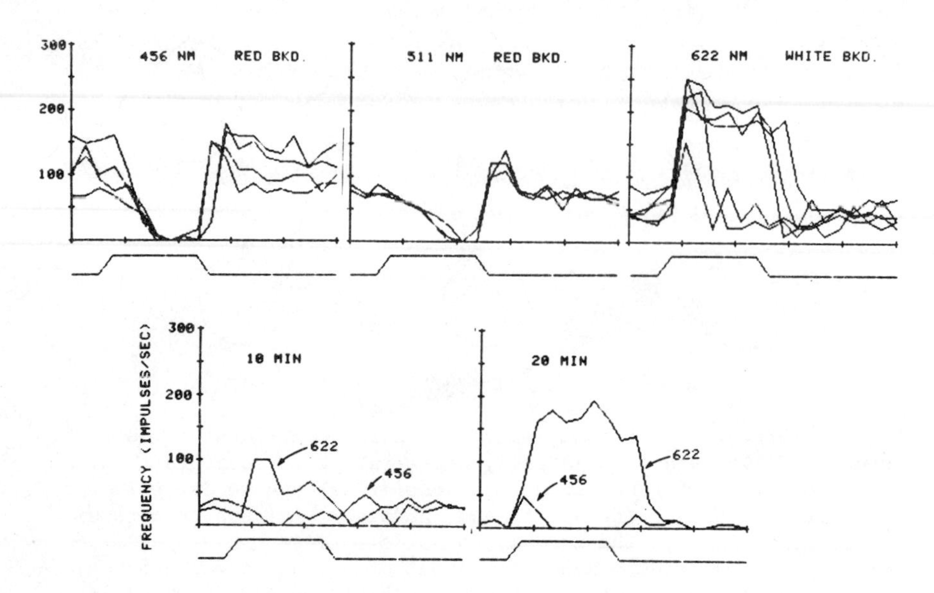

Fig. 3. Responses of a Type II-like L/M opponent ganglion cell subserving the fovea of the rhesus monkey to three narrow band stimuli 456nm (blue), 511nm (green) and 622 nm(red). The upper set of responses are obtained with a 20° stimulus; the lower ones with a 10 and 20 minute diameter spot centered on the cell's receptive field. This cell is excited by red and inhibited by green and blue lights. The records above show responses to different energies of stimulation at the same wavelength. Those to red extend from supra-saturating to just above threshold levels. The former can be identified by a prolongation of the depolarization at off and the higher impulse frequency before the stimulus occurs. This cell shows opponency between L and M cones because the sensitivity to the 511 nm stimulus for inhibition in the presence of a red adapting field is more than 10 times greater than that to 456nm. For the S cone mechanism 456nm would be more than 100 times more sensitive than 511 nm under the same conditions. The responses below show that there is some color opponency to the smallest test spots but the small red spot is more effective than the blue spot and its excitatory response is quicker than the inhibitory response to the shorter wavelengths (456nm and 511nm). The abscissa scale is in 133 msec intervals; the duration of the stimulus is below each set of responses.

There are reports, however, of units with red/green, presumably L/M, opponent interactions that have a Type II receptive field organization in the primate retinogeniculate system (Wiesel and Hubel, 1966; DeMonasterio and Gouras, 1975; Dreher, Fukada and

Rodieck, 1976; Krüger, 1977). All are in agreement that such cells are rare in comparison with Type I L/M opponent cells in which a center/surround color opponent receptive field organization can be delineated. It is not easy to distinguish a Type II from a Type I L/M cell when receptive field sizes are small. I believe it is possible to make a Type II-like cell have Type I properties when stimuli are small enough (Gouras and Zrenner, 1981a). Figure 3 shows temporal differences in the center and surround responses of a Type II-like retinal ganglion cell subserving the foveal area of the rhesus monkey. This cell receives an inhibitory input from M cones and an excitatory input from L cones. The inhibitory surround response is slower than the excitatory center response. In this paper I assume that all L/M opponent cells are Type I cells.

The best candidates for Type II retinogeniculate cells, are those subserving the S cone mechanism. These cells have a larger receptive field size than Type I L/M cells. I have seen this consistently (Gouras, 1968; DeMonasterio and Gouras, 1975; Gouras and Zrenner, 1981) and it has also become apparent indirectly with the use of a moving edge of brightness and/or spectral contrast as a stimulus (Gouras and Eggers, 1983). The cells mediating the S cones seem to be homogeneous group, all with virtually identical properties: a relatively large receptive field size, a similar and strong opponent input from the long wave, presumably L and M cones, and usually, perhaps always, their S cone input is excitatory. These cells comprise a small fraction about 5% of the cells in the retinogeniculate system (Gouras, 1984).

Figure 4 illustrates how a mosaic of L, M and S cones in primate is subserved by parallel systems of retinogeniculate cells. The cells with the smallest receptive field sizes, which near the fovea are extremely small and probably include the midget ganglion cell system, are Type I cells in which one class of L or M cone(s) forms the center and the other class forms the antagonistic surround. These cells can be either on- or off-center. There is a considerable variation in the degree of L/M opponency these cells show. This heterogeneity has led to different classification schemes (Creutzfeldt, Lee and Elephandt, 1979) but the overall picture seems similar among most laboratories. I have found that an effective way to go from color opponent responses to cone mechanisms is by the use of selective chromatic adaptation in conjunction with action spectra (Fig. 3, legend). What I consider of maximal importance in this cell system is the relatively small sizes of the receptive fields of these cells and the presence in most cases of some degree of L and M cone opponency.

The third major channel of retinogeniculate cells, called Type III cells in Fig. 4, are functionally distinct in the following ways. They do not show overt L/M opponency because both L and M cones appear to subserve both center and surround mechanisms. They receive no input from S cones, have larger receptive field centers, produce phasic responses to maintained stimuli, have relatively

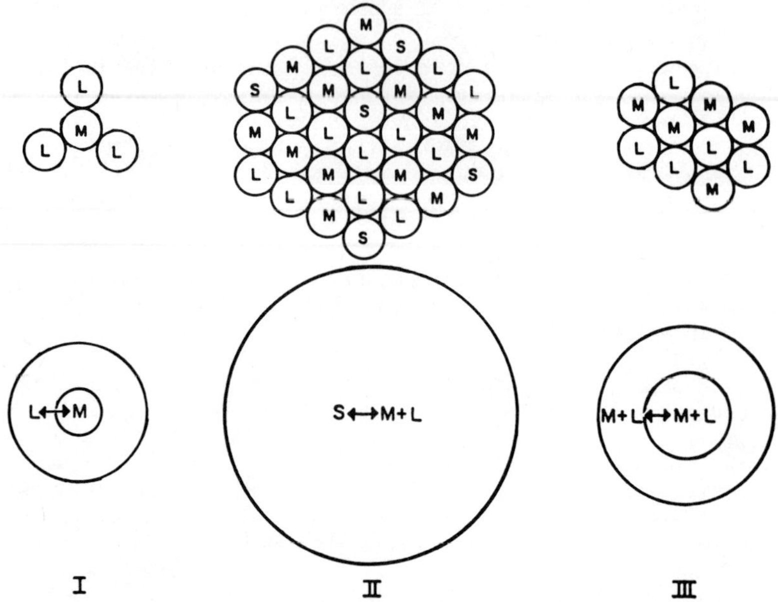

Fig. 4. A scheme of how parallel systems of retinogeniculate neurons subserve the same cone photoreceptor mosaic. Each small circle in the mosaic signifies a single cone. There are equal numbers of M and L but fewer S cones. An S/M+L ratio of 0.17 is used but 0.1 is probably a better estimate (Gouras, 1984). Type I units have an input from only one type of cone in the center and another cone type in the surround. Type II units receive antagonistic (S versus M+L) signals from a quasi-coextensive population of cones; their receptive fields are relatively large. Type III cells have a concentrically organized receptive field organization but with both center and surround mechanisms receiving signals from M and L cones.

fast conduction velocities and are relatively more concentrated in peripheral retina.

There is considerable agreement that these three functionally distinct sets of retinogeniculate cells subserve each area of primate retina in parallel. There is some question whether additional types exist. There are cells which resemble Type I L/M cells but do not show overt L/M opponency. There are some Type III phasic cells which are preferentially inhibited by long wave stimuli, Type IV cells of Wiesel and Hubel (1966). We have suggested that the excitatory discharge to green but not red lights in Type IV cells may be due to a maintained rod signal (Gouras and Eggers, 1983). Undoubtedly the three major classes of cells shown in Fig. 4 are an oversimplification but these categories cover most cells and provide a useful scheme for interpreting the layering in

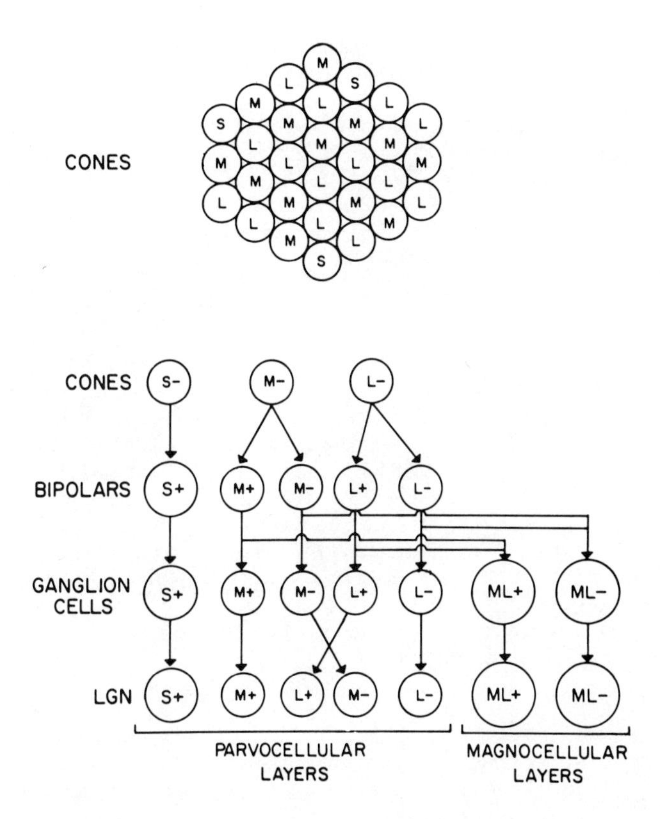

Fig. 5 A scheme showing the serial processing of receptive field center signals of parallel retinogeniculate channels subserving the same cone mosaic. The S cone system is considered as on-center cells for S cones even though it is thought to have a Type II receptive field and could as well be off-center for M+L cones. The L and M system is subserved by two different subsystems, each with its own on- and off-center cells; one is composed of tonic Type I, mostly color opponent cells; the other is composed of phasic type III non-color opponent cells.

the lateral geniculate nucleus.

Figure 5 shows how a mosaic of cones in primate retina are subserved by a serial system of on- and off-center neurons for each of these three parallel systems in the retinogeniculate pathway. The cones, themselves, are off-center neurons. From the bipolar level onward there are separate subsystems of on- and off-center cells subserving each system. No off-center cells have been relegated to the S cone channel because off-responses are rare from this system. The phasic or Y-like magnocellular channel is shown for simplicity to be sharing the same bipolars as the tonic or

X-like parvocellular channels; we suspect that they use a separate set of bipolars. The major transformation occurring between the retinal output and the geniculate is the segregation of different functional classes of cells to different geniculate lamina. The major segregation occurring in the lateral geniculate nucleus, second only to ocularity segregation, is that in which tonic or X-like and usually color opponent cells, go to the parvocellular layers and phasic or Y-like non-color opponent cells to to the magnocellular layers (Gouras and Zrenner, 1981b; Gouras, 1984). On center cells are concentrated in the dorsal and off-center ones in the ventral parvocellular layers (Schiller and Malpeli, 1978).

It is important to re-evaluate the traditional view of the retinogeniculate pathway in relation to this scheme. After the discovery of color opponent interactions in the retina and the lateral geniculate nucleus the idea emerged that the primate retinogeniculate pathway contained cells which might be the counterparts of Hering's three color opponent channels. Those cells which were excited by red and inhibited by green light or vice versa were considered to be the red/green opponent channel of Hering. Those excited by blue and inhibited by yellow or vice versa were considered to be the blue/yellow opponent channel of Hering. Those cells which were excited and those which were inhibited by all spectral lights were considered to be white and black opponent channels of Hering.

In order to test this hypothesis we have been using a moving edge as a stimulus in which spectral (chromatic) contrast can be separated from effective energy (brightness) contrast (Gouras and Eggers, 1982,1983, 1984). The results show that cells excited or inhibited by all spectral lights, usually Type III phasic cells, cannot distinguish white from yellow; they can only distinguish a brightness contrast across a contour and this is detected <u>only</u> by the M and L cones. The only retinogeniculate cells that can distinguish white from yellow are those subserving the S cone mechanism; they are excited by white and inhibited by yellow at <u>all</u> brightness contrasts across a moving edge. Therefore, the so called blue/yellow channel plays the pivotal role in distinguishing white as well as blue from yellow. No <u>one cell</u> in the primate retinogeniculate channel can distinguish the color of a stimulus. The message for color at this point is contained among but not within a single cell.

An identical situation occurs in the so called red/green opponent cells, composed of the Type I L/M opponent system; this is the most ubiquitous cell system in the primate retinogeniculate pathway, especially in the fovea. Figure 6 demonstrates the ambiguity of the color information contained in the responses of two Type I cells, an on-center M+/L- and an on-center L+/M- cell. Both of these cells respond strongly to a small white spot flashed on the center of each cell's receptive field. An identical small green or red spot will produce a weaker response than white (or

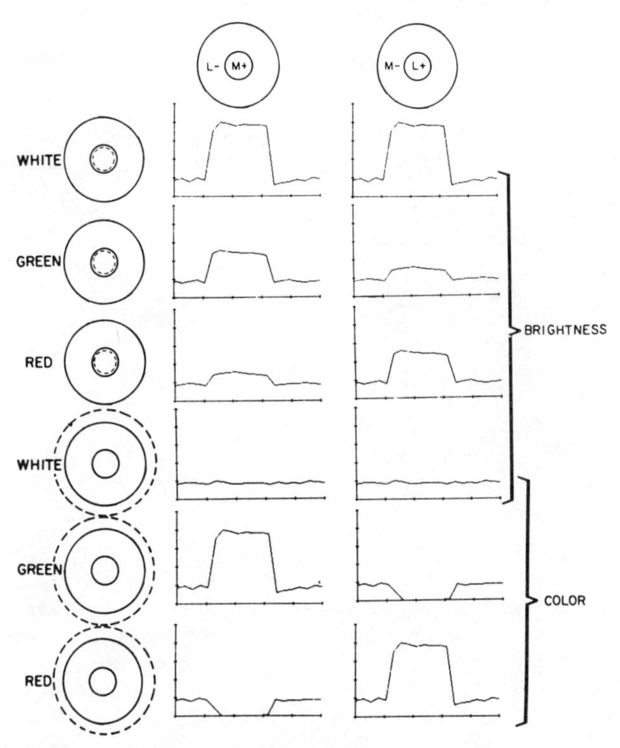

Fig. 6. This illustrates how color coding information is contained not in a single L/M opponent cell but in two such cells of opposite color opponent polarity. One cell has an M on-center/L off-surround; the other has an L on-center/M off-surround. The column on the left shows how stimuli of various colored spots (dashed circles) subtend the receptive field of the cells. Small spots which lie entirely within the receptive field center are shown above; large spots which extend beyond the antagonistic surround are shown below. The responses to each of the spatially and spectrally unique stimuli are shown in a column below the appropriate cell. These responses are modelled after a typical Type I L/M opponent cell. The information about the color of the stimulus is ambiguous in the response of any one cell but not if both cells are considered simultaneously. It is also apparent that information for color coding is lost as stimulus size is reduced. A similar effect may occur in the time domain (Gouras and Zrenner, 1978).

yellow) but the green spot will be more effective for the M+/L- than for the L+/M- cell; the red spot will show the reverse behavior. Neither cell can distinguish a small dim white spot from a red or a green spot. A large white spot will produce little to no response from either cell because it strongly stimulates both center and surround mechanisms. A large green spot will excite the M+/L- cell strongly and inhibit the L+/M cell; the large red spot will do the reverse. Again the discharge frequency of any one of these cells provides an ambiguous message about the color of a stimulus. Such a cell is therefore not equivalent to any of Hering's three opponent channels. The color information is contained in the network and not in any one cell at this point in the central nervous system. To decode this signal one must examine several cells simultaneously. Visual cortex must have such a decoding circuit.

Figure 7, left, illustrates such a hypothetical circuit. It uses several on-center M+/L- and L+/M- geniculate cells to abstract simultaneous M/L cone contrast in what has been christened a double color opponent cell by Nigel Daw (1968). This circuit combines in the center of its receptive field excitation from on-center M+/L- cells with inhibition from L+/M- ones; the surround combines the converse input. A requirement for spatial integration over at least several M and L cones both in the center and the surround of its receptive field prevents it from responding to small white spots. This cell responds best when the difference between M and L cone activation across a contour is maximal, i.e. a green spot in a red background for a concentrically organized cell. Similarly a double opponent cell responsive to a maximum contrast for the L cone mechanism can be formed by reversing the polarity of all synapses. These simultaneous color contrast cells sacrifice the maximal spatial resolution of the cone mosaic in order to detect maximum hue contrast. In order to exploit the maximum spatial resolution of the receptor mosaic, brightness contrast detectors can be formed in parallel using the same geniculate input. Figure 7, right, shows how a brightness contrast detector can be formed using only one cone in the center and another in the surround, the ultimate spatial contrast. Either M or L cones can be used since, their spectral ranges are relatively similar. Brightness contrast has therefore a potentially finer spatial resolution than hue contrast. It is noteworthy that hue contrast detectors can be formed using only excitatory on-center units on both sides of a contour whereas brightness contrast detectors cannot. The predominance of on-center responses in the S cone system may reflect its relatively strong input to chromatic rather than brightness contrast.

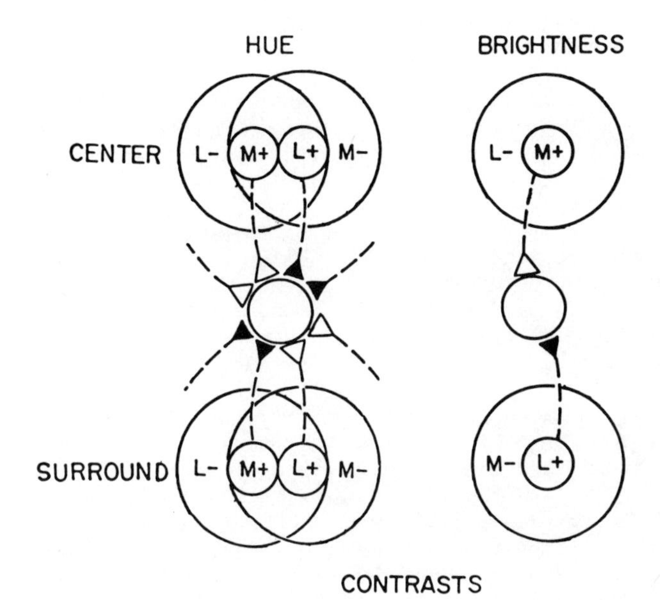

Fig. 7. A circuit on the left designed to detect simultaneous color (hue) contrast signals in striate cortex from an array of Type I L/M color opponent geniculate cells subserving neighboring cones. The cortical unit (open circle) receives an excitatory input from M and an inhibitory input from L on-center geniculate cells in its receptive field center; it receives a similar input but with opposite synaptic polarity from its surround. A response from either the center or the surround mechanism requires the activation of at least two or more M and L on-center units in the surround or the center of the receptive field; this prevents the cell from responding to small white spots. The cell responds best to green in the center and red in the surround of its receptive field, i.e. hue rather than brightness contrast. The same geniculate input can form brightness contrast detectors with the circuit shown on the right. This can be formed with only one cone, either M or L, in the center, and a neighboring cone, either M or L, in the surround via midget retinogeniculate Type I L/M opponent units.

L/M double opponent units have been found in visual cortex by many investigators (Hubel and Wiesel, 1968; Michael, 1978; Gouras and Krüger, 1979; Livingstone and Hubel, 1984). There is evidence that brightness contrast detectors receive a color opponent input (Krüger and Gouras, 1980), which is another requirement of this model. The model predicts that the receptive field sizes of double opponent L/M units are larger than those of brightness contrast detectors.

A similar strategy can be used to form double opponent units for S/M+L cone contrast, i.e. blue/yellow contrasts, but experimental evidence for their existence is lacking. Here the circuitry is made easier if, as proposed, the S cone system is composed of Type II cells which do not contribute to fine spatial resolution by brightness contrast detection.

Figure 8 shows a general scheme for building all the major forms of color contrast including brightness, hue and saturation detectors from the retinogeniculate system. Concentrically organized units can be used to form simple and complex slit and bar detectors; non-concentrically organized ones seem necessary to form edge detectors. The most difficult detector to construct is the white/black or "desaturation" detector, which seems to require some form of logical integrator (Gouras, 1985). The crux of this model is that each color contrast detector operates relatively independently of the other in distinguishing a visual contour. In this model, movement across visual contours is essential to vision and events that do not produce time varying spatial contrast on the receptor mosaic become invisible. There are many possibilities for preserving color constancy in this system for large variations in overall brightness or, for brightness gradients, so common in most visual scenes. One such mechanism would be to have excited hue contrast detectors veto brightness contrast detectors since the former are more reliable indicators of objects in the natural world (Gouras, 1984).

A major consideration is how the array of different contrast detectors unite to form a colored object. It seems essential for this model that spatial polarity of each detector be established and limited to similar detectors across large areas of visual space. An interesting point in this analysis is that any one hypothetical Hering opponent channel is still not a single neuron but pairs of double opponent contour detectors of opposite polarity.

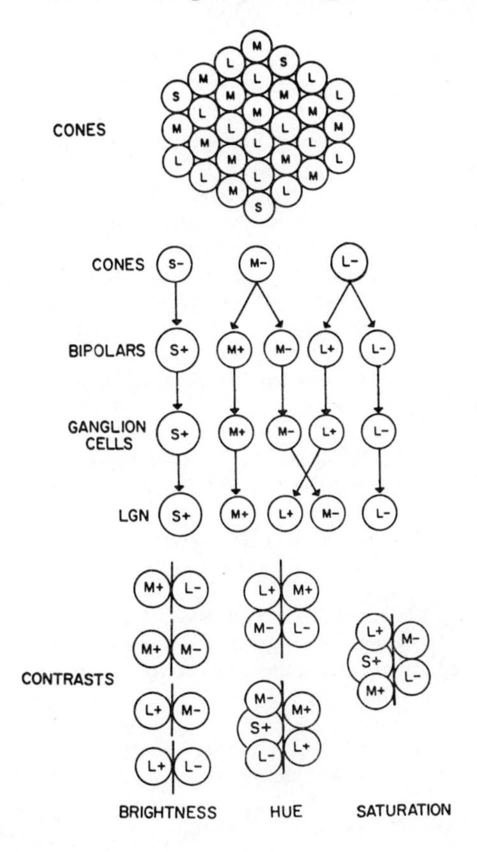

Fig. 8. A scheme illustrating how cortical brightness, hue (red/green and yellow/blue) and saturation (white/black) contrast detectors are formed from the receptive field centers of retino-geniculate cells subserving the same cone mosaic in primate retina. These detectors sense contrast across neighboring cones. Brightness contrast detectors have either M or L on-center units on one side side of a contour. They have the finest spatial resolution. Red/green contrast detectors have M or L on-center units on opposite sides of a contour in addition to the subcircuitry in Fig. 7, right. Yellow/blue contrast detectors have S on-center on one side and both M and L on-center units on opposite sides of a contour; these cells also require the spatial integration requirement of Fig. 7, right. The third variable of color vision, saturation is sensed by a "desaturation" (white/black) contrast detector which responds only if all three S, M, and L on-center units are active on one side of a contour.

References

Creutzfeldt, O.D, Lee, B.B., Elepfandt, A. (1979). A quantitative study of chromatic organization and receptive fields of cells in the lateral geniculate body of the rhesus monkey. Exp. Brain Res. 35, 527-545.

Dartnall, H.J.A., Bowmaker, J.K. and Mollon, J.D. (1983). Microspectrophotometry of human photoreceptors. In Color Vision: Physiology and Psychophysics (eds. J.D. Mollon and L.T. Sharpe) Academic Press, London.

Daw, N. (1968). Colour-coded ganglion cells in the goldfish retina: Extension of their receptive fields by means of new stimuli. J. Physiol 197, 567-592.

DeMonasterio, F.M., Gouras, P. (1975). Functional properties of ganglion cells of the rhesus monkey. J. Physiol. 251, 167-196.

DeValois, R.L. (1960). Color vision mechanisms in the monkey. J. Gen Physiol. 43, 115-128.

Dreher, B., Fukada, Y., Rodieck, R. (1976). Identification classification and anatomical segregation of cells with X-like and Y-like properties in the lateral geniculate nucleus of Old World primates. J. Physiol., Lond 258, 433-452.

Gouras, P. (1968). Identification of cone mechanisms in monkey ganglion cells. J. Physiol. 199, 533-547.

Gouras, P. (1984). Colour Vision. In Progress In Retinal Research. (eds. G.J. Chader and N.N. Osborne) Pergamon Press, Oxford.

Gouras, P. (1985). Parallel processing of colour contrast in visual cortex. In Models of the Visual Cortex. (eds. A. Rose and V. Dobson) Wiley and Sons, USA.

Gouras, P., Eggers, H.M. (1982). Ganglion cells mediating the signals of blue sensitive cones in primate retina detect white-yellow borders independently of brightness. Vis. Res. 22, 675-679.

Gouras P., Eggers, H.M. (1983). Responses of primate retinal ganglion cells to moving spectral contrast. Vis. Res. 23, 1175-1182.

Gouras, P., Krüger, J. (1979). Responses of cells in foveal visual cortex of the monkey to pure color contrast. J. Neurophysiol 42, 850-860.

Gouras, P., Zrenner, E. (1978). Enhancement of luminance flicker by color-opponent mechansims. Science 205, 587-589.

Gouras, P., Zrenner, E. (1979). The blue sensitive cone system. Excerpta Medica Int. Cong. Ser. 450/1, 379-384.

Gouras, P., Zrenner, E. (1981a). Color coding in the primate retina. Vis. Res. 21, 1591-1598.

Gouras, P., Zrenner, E. (1981b). Color Vision: A review from a neurophysiological perspective. Prog. Sens. Physiol 1, 139-179.

Hubel, D.H., Wiesel, T.N. (1960). Receptive fields of optic nerve fibres in the spider monkey. J. Physiol, London 154, 572-580.

Krüger, J. (1977. Stimulus dependent color specificity of monkey lateral geniculate neurons. Exp. Brain Res. 30, 297-311.

Krüger, J., Gouras, P. (1980). Spectral selectivity of cells and its dependence on slit length in monkey visual cortex. J. Neurophysiol. 43, 1055-1069.

Livingstone, M.S., Hubel, D.H. (1984). Anatomy and physiology of a color system in the primate visual cortex. J. Neurosci. 4, 309-356.

Malpeli, J.G., Schiller, P.H. (1978). Lack of blue Off-center cells in the visual system of the monkey. Brain Res. 141, 385-389.

Michael, C.R. (1978). Color vision mechanisms in monkey striate cortex: dual-opponent cells with concentric receptive fields. J. Neurophysiol. 41, 572-588.

Schiller, P.H., Malpeli, J.G. (1978). Functional specificity of lateral geniculate nucleus laminae of the rhesus monkey. J. Neurophysiol. 41, 788-797.

Svaetichin, G. (1956). Spectral response curves from single cones. Acta physiol. scand. 134 suppl. 19-46.

Wiesel, T.N., Hubel, D (1966). Spatial and chromatic interactions in the lateral geniculate body of the rhesus monkey. J. Neurophysiol. 29, 1115-1156.

Supported by research grants EY02591, EY04138 from the National Eye Institute, NIH, a Center Grant from the National Retinitis Pigmentosa Foundation, The House of St. Giles the Cripple and Alcon Laboratories

Non-oriented Double Opponent Colour Cells are Concentrated in Two Subdivisions of Cortical Layer IV*

C.R. Michael

Department of Physiology, Yale Medical School, 333 Cedar Street, New Haven, Connecticut 06510, USA

Microelectrode studies of single cells in the monkey's visual pathway have revealed that visual information is processed by opponent mechanisms. In the case of luminance contrast, individual neurons receive signals from two sets of identical receptors, one group occupying a circular area and the other surrounding it in an annular fashion. Because of the specific excitatory and inhibitory synaptic connections, illuminating the center of the cell's receptive field either increases or decreases its firing rate while stimulating the surround has the opposite effect. Illumination of the entire receptive field leads to mutual antagonism and thus this opponent spatial organization discriminates against diffuse light in favor of contrast.

Color is also integrated through an opponent mechanism. Cells are connected with two sets of cones containing different visual pigments and thus having different spectral sensitivities. As a result of the synaptic connections, one type of cone excites the cell and the other inhibits it. Consequently, the neuron increases its rate of firing to one set of wavelengths and decreases it to another set, thus discriminating against white light.

In the monkey's visual system both opponent color and opponent contrast integrative mechanisms converge on single retinal ganglion cells and lateral geniculate neurons. In the retina the majority of the color coded ganglion cells have a circular excitatory (on) or inhibitory (off) field center connected with one cone type and an annular surround of the opposite sign connected with a different class of cone (Gouras, 1968). At the geniculate level the organization appears to be basically unchanged, with the majority of the cells being of this same type (Wiesel and Hubel, 1966).

What happens to the color information carried by the geniculate axons to the striate cortex? For instance, one important

*Supported by National Eye Institute Grant EY 00568.

issue to be resolved is whether color and spatial contrast information eventually occupy the same channels or are kept separate in different groups of opponent color and opponent contrast cortical cells. Several years ago I began to study the properties of color sensitive cells in the monkey's striate cortex. I found a variety of cell types which were processing color and color contrast aspects of the visual image. I will concentrate here on just the first stage of this integrative pathway. I have been re-examining the physiological properties and laminar location of these first order cells in a recent study of cortical layer IV.

The cortical cells which I believe are receiving direct geniculate inputs have concentric receptive fields with one red/green opponent color system in the field center and the reverse opponent color organization in the annular surround (Michael, 1978). A cell's field thus consisted of a red on, green off center and a green on, red off surround or of the opposite arrangement. In contrast with the fact that most cortical cells were orientation sensitive and binocular, the concentric cells had non-oriented

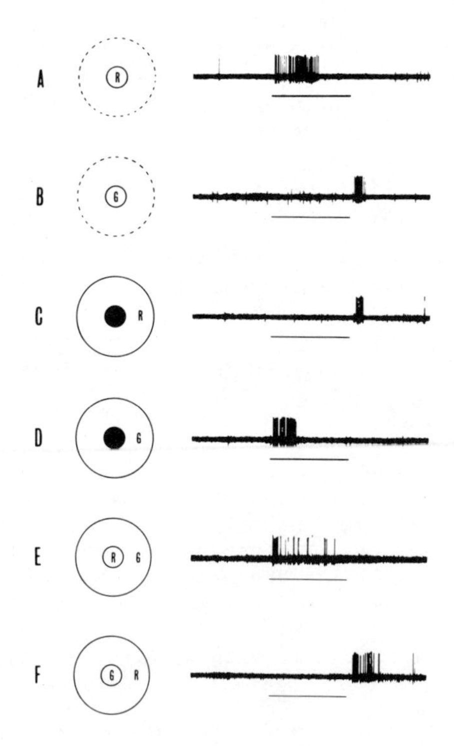

Fig. 1. Responses of a double opponent color concentric cortical cell to red (630nm) and green (500nm) spots and annuli (adapted from Figures 1 and 2, Michael, 1978).

fields and were always driven by only one eye. Figure 1 illustrates the responses of one of these concentric cells to monochromatic spots and annuli. Red and green spots in the field center produced on and off responses, respectively, while red and green annuli had the reverse effects. Diffuse monochromatic or white light in any form produced no discharge. A red spot encircled by a green annulus or a green spot surrounded by a red annulus elicited the strongest on or off responses from the cell. In other words, these dual opponent cells were most sensitive to the simultaneous presentation of two different colors, one covering the field center and the other illuminating the surround. They are thus likely to be involved in the perception of simultaneous color contrast phenomena. In such situations two colored objects positioned adjacent to each other appear more saturated than they do when placed apart. A neural correlate of this effect can be seen in the strong discharge elicited in the last two examples of Figure 1.

The spectral sensitivity of the field center of this same cell is shown in Figure 2. In the light adapted state there are two curves, one associated with the long wavelength on response and the other with the short wavelength off discharge. The neutral point at about 540nm marks the wavelength at which excitation was balanced by inhibition and consequently there was no response. The points connected by interrupted lines represent the spectral sensitivities obtained under two different monochromatic background

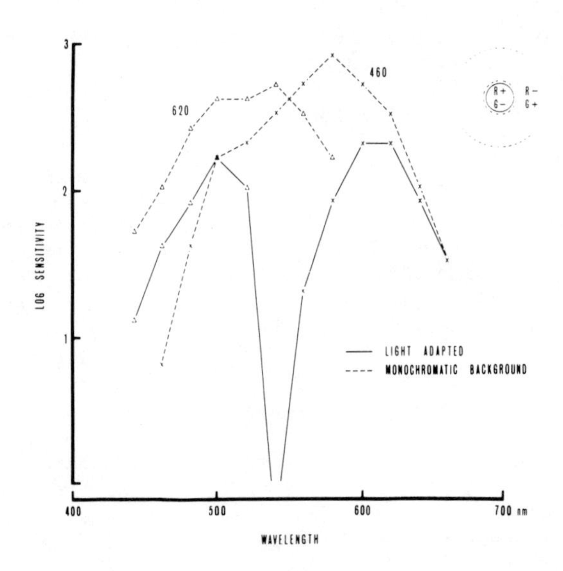

Fig. 2. Spectral sensitivity of the red on, green off center of the cell in Figure 1, under conditions of white or monochromatic backgrounds (Figure 3, Michael, 1978).

conditions. It is apparent that the on component was mediated by red sensitive cones and that the off system was under the control of green sensitive cones.

When illumination of the surrounds of these cells evoked discharges, the spectral sensitivity curves were the same as those in Figure 2, with the response signs reversed. However, for many of the concentric cells it was impossible to elicit an independent discharge from the surround with monochromatic annuli. Nevertheless the presence of a "silent" antagonistic surround could always be demonstrated by the failure of a large colored spot to produce any effect. Area sensitivity curves invariably revealed that peripheral suppression was present for both phases of the center opponent color system.

Recently I have undertaken to extend and refine my earlier observations regarding the distribution of the various color cell types in the different layers of the primate striate cortex and, in particular, to examine in detail the location of the double opponent non-oriented concentric cells in the various subdivisions of layer IV. The results indicate that the non-oriented concentric cells with double opponent color properties were strictly confined to layers IVA and IVC_b; this cell type was also found in the lower half of IVC_a but there it was always of the broad band class. Concentric cells of any type were absent from IVB. Simple cells were found in significant numbers in IVB and upper IVC_a and were always broad band while those few found in IVA and IVC_b were usually double opponent color but sometimes were broad band. Color-sensitive complex and hypercomplex cells were found in the supragranular and infragranular layers (laminae 2, 3, 5 and 6). Double opponent concentric cells were occasionally seen in the supragranular layers but never in the infragranular laminae.

Figure 3 is a reconstruction of an oblique electrode track through all of layer IV. A total of thirty-nine cells were studied within the boundaries of lamina IV. The unit marked by the lesion L_1 was the first non-oriented concentric cell encountered in this track and was a reliable sign that the electrode had entered layer IVA. Eight of the nine cells in IVA had concentric center/surround fields and two clusters of these neurons had double opponent color properties. In contrast, in layer IVB, which gets its input from IVC_a (Blasdel et al., 1983a,b; Lund et al., 1983), all of the cells studied were orientation sensitive and none were color coded. In the upper half of IVC_a the cells were broad band and orientation selective while in the bottom half they were broad band center/surround cells. In the last portion of the track, through IVC_b, a large fraction of the cells were color selective and had non-oriented center/surround concentric receptive fields.

In highly tangential penetrations the segregation of the cell types in layer IV becomes even more obvious. For example, in Figure 4 a highly oblique electrode traveled through layer IVC_b

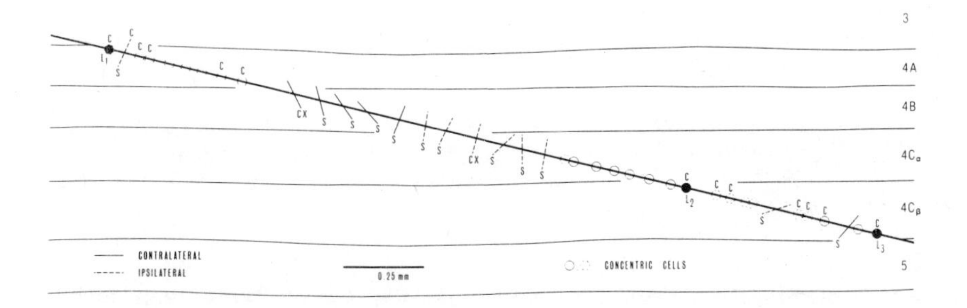

Fig. 3. Oblique electrode track through all of the subdivisions of layer IV. The circles indicate the locations of non-oriented concentric cells. The letter C above a cell's position signifies that it is specifically color sensitive.

for over two millimeters before the track was terminated. The lesion at L_2 marks the first non-oriented concentric color cell seen after a long sequence of broad band simple and concentric cells; the histology revealed that the electrode had just entered layer IVC_b. An electrode passing through IVC_b detected little or no trace of orientation preference in the unresolved background activity because, as shown in this track, most of the cells were concentric neurons with only an occasional orientation sensitive simple cell interspersed along the penetration. There were five clusters of color cells along the track. The members of a cluster always shared common response signs and spectral sensitivities. For example, the neurons in the first group were all double opponent cells with red on, green off centers and the reverse surrounds. The next cluster had the opposite double opponent organization; the remaining groups alternated in their color properties.

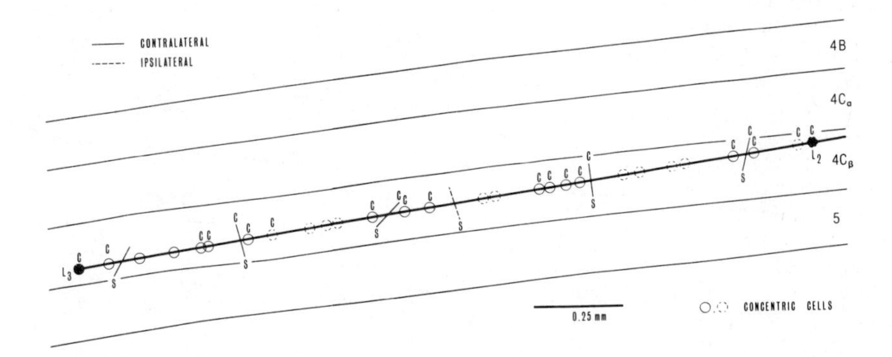

Fig. 4. Tangential electrode track through layer IVC_b, revealing a high number of concentric and color specific cells.

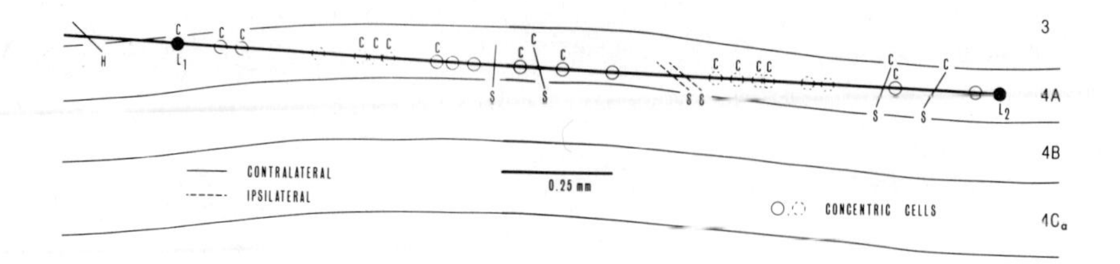

Fig. 5. Highly oblique track through layer IVA. Most of the cells were concentric and about half were double opponent.

Figure 5 illustrates another oblique electrode track, in this case through layer IVA. From the cortical surface to the cell marked by lesion L_1 the neurons were all orientation selective. The cell at L_1 was a red on, green off center non-oriented double opponent unit which was found at the top of IVA. In this penetration there were five clusters of color cells and, as in IVC_b, the units in each group had common spectral characteristics and center sign responses. In both layers IVC_b and IVA a large majority of the cells were of the concentric center/surround type and the remaining few were simple neurons (Hawken and Parker, 1984). Many of the cells in both laminae were color coded. In other words, the tracks through IVA and IVC_b seemed very similar. These two sublayers apparently receive separate sets of parvocellular geniculate fibers (Blasdel and Lund, 1983) but it remains to be demonstrated that there is any difference between the cells in IVA and those in IVC_b.

Consistent with the exclusive innervation of IVC_a by broad band magnocellular fibers (Hubel and Wiesel, 1972; Blasdel and

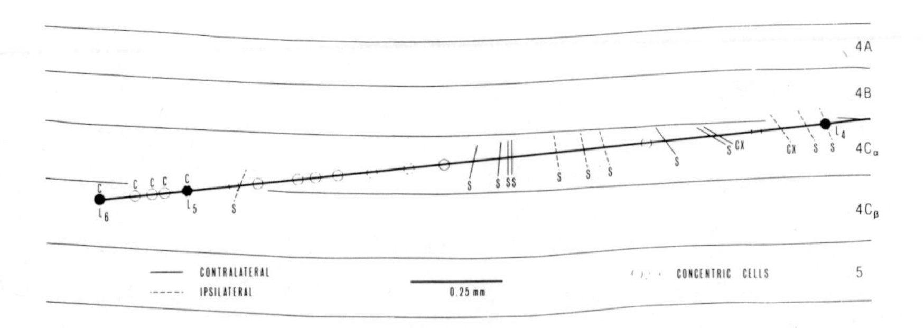

Fig. 6. Low angle electrode penetration through layer IVC_a. All cells in IVC_a were broad band. Units in the top half were simple while those in the bottom half were concentric.

Lund, 1983), color cells were extremely rare between layers IVA and IVC_b. For instance, Figure 6 illustrates the types of cells encountered in an electrode track which passed through layer IVC_a. There were no opponent color cells; all of the units had broad band spectral sensitivities and high contrast sensitivities. In this track most of the cells in the top half of the layer were simple units while in the bottom half, all of the neurons, except one, were concentric cells. These results are typical for a penetration through IVC_a: a fairly strict segregation in which most of the simple cells were confined to upper IVC_a while lower IVC_a contained concentric cells lacking orientation selectivity. Near the end of the track the electrode entered layer IVC_b and passed through a cluster of double opponent concentric neurons, the first color cells seen in over three millimeters. It was always possible to

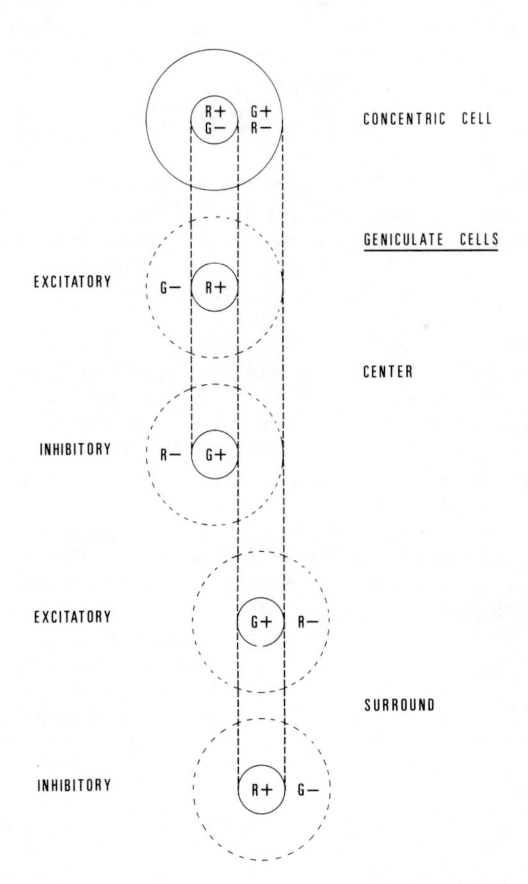

Fig. 7. Proposed synaptic mechanism to explain the response properties and receptive field organization of the double opponent color non-oriented concentric cells.

determine when an electrode had left IVC_a and entered IVC_b, in spite of the predominance of concentric cells both in the bottom half of IVC_a and throughout IVC_b. In every case the center size of fields became smaller, the responses changed from transient to sustained, contrast sensitivity decreased and color selectivity appeared (Blasdel and Lund, 1983; Blasdel and Fitzpatrick, 1984; Hawken and Parker, 1984).

Given that the concentric dual opponent cells are highly concentrated in the two cortical sublaminae which receive heavy parvocellular afferent inputs and are absent from the sublayer in which the magnocellular axons end, it seems reasonable to assume that these cells are the first stage in the cortical integration of color contrast information. They presumably receive direct synaptic contacts from the axons of the most common opponent color geniculate cells, i.e. the ones with a center/surround receptive field and red/ green opponency. Figure 7 illustrates a possible mechanism to explain the development of double opponency in the cortex. A single red on center, green off surround geniculate axon makes excitatory synapses on the concentric cortical cell while a green on, red off unit makes inhibitory contacts on it (presumably through an interneuron). The field centers of these two afferents are coincident with that of the concentric cell. A red spot in the concentric cell's field center excites the first geniculate cell and, in turn, the cortical neuron. A green spot excites the second geniculate afferent but, because of the inhibitory synapse, suppresses the firing of the concentric unit. Thus one has a red on, green off field center for the cortical cell.

The surround is probably made up from the field centers of a large number of red on and green on center geniculate fibers. Here the synaptic connections are the reverse of those in the cortical cell's field center; green on center geniculate units are excitatory and red on center are inhibitory. Thus a green annulus increases the cortical cell's firing while a red one prevents it. As a consequence of these connections, simultaneous presentation of one color in the field center and the complementary color in the surround would evoke maximal excitatory (on) or inhibitory (off) responses. On the other hand, diffuse colored light, which is often a reasonable stimulus for geniculate cells, is totally ineffective for the concentric cell. The explanation for this effect is that a large red spot has an excitatory influence in the field center, via the red on center geniculate axon, and an inhibitory one in the surround, through the other red on center axon. The two effects cancel one another. A similar argument holds for a large green spot.

Livingstone and Hubel (1984) have recently reported that non-oriented double opponent color cells are present within the cytochrome oxidase blobs of layers II and III of the monkey's striate cortex. Since I did not find such cells above layer IV in my earlier experiments on the rhesus monkey (Michael, 1978), I have re-

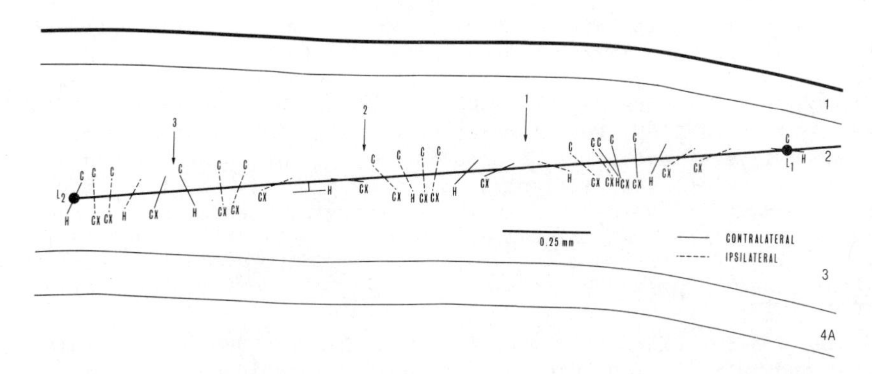

Fig. 8. Oblique electrode track through layers II and III. Only orientation sensitive complex and hypercomplex cells were seen.

examined the supragranular layers of the striate cortex of the cynomolgus monkey for these neurons. However, many of the penetrations through layers II and III looked like Figure 8, with all of the cells encountered being orientation sensitive. The cells were exclusively complex or end-stopped hypercomplex neurons with color being a necessary stimulus parameter for about half of the cells. There were no simple or concentric units in this penetration, which did not reach layer IV. The color specific complex and hypercomplex cells, which were poorly influenced, if at all, by white light, occurred in discrete clusters which had common color properties. The presence or absence of color specificity appeared to bear no relationship to the orderly rotation of the receptive field axes seen between lesion L_1 and arrow 3. Assuming that the blobs contain non-oriented color cells but that they are relatively far apart, the track in Figure 8 may not have passed through any of them or perhaps it grazed them tangentially where both color and orientation might be represented.

Regardless of the type of color cells found in the supragranular layers, it seems unlikely that they receive any significant number of direct contacts from opponent color geniculate axons. Instead they must get their afferent inputs from cells within the cortex itself, namely from the color cells in IVA and IVC_b which probably project to layer III (Lund and Boothe, 1975). For instance, it has recently been shown (Blasdel et al., 1983a,b; Lund et al., 1983) that spiny stellate cells in the upper half of IVC_b project to lower layer III and thus they could be the double opponent afferents to supragranular color cells. The non-oriented concentric color cells seen in layer III by Livingstone and Hubel (1984) may actually receive some direct parvocellular inputs from fibers whose terminals end mostly in IVA but send some collaterals up into layer III (Blasdel and Lund, 1983; Lund et al., 1983). It seems quite unlikely, however, that the recently described intercalated geniculate cell input to layer III of the squirrel monkey's striate cortex (Fitzpatrick et al., 1983) plays much of a role, if any, in supragranular color processing. In the rhesus and cynomolgus monkeys these geniculate layers represent the peripheral but not the central part of the visual field, where color vision is so strongly developed. Furthermore, no one has shown that the intercalated layers contain color opponent cells. The non-oriented color cells in layers III and II are thus probably receiving inputs from concentric cells in layer IV and therefore represent the second stage of cortical color processing.

References

Blasdel, G.G., Fitzpatrick, D. and Lund, J.S. (1983a). Physiological and anatomical studies of retinotopic maps in macaque striate cortex. Invest. Ophthal. Vis. Sci., 24, 266.

Blasdel, G.G., Fitzpatrick, D. and Lund, J.S. (1983b). Organization and intracortical connectivity of layer 4 in macaque striate cortex. Soc. Neurosci. Abstr., 9, 476.

Blasdel, G.G. and Lund, J.S. (1983). Termination of afferent axons in macaque striate cortex. J. Neurosci., 3, 1389-1413.

Blasdel, G.G. and Fitzpatrick, D. (1984). Physiological organization of layer 4 in macaque striate cortex. J. Neurosci., 4, 880-895.

Fitzpatrick, D., Itoh, K. and Diamond, I.T. (1983). The laminar organization of the lateral geniculate body and the striate cortex in the squirrel monkey (Saimiri sciureus). J. Neurosci., 3, 673-702.

Gouras, P. (1968). Identification of cone mechanisms in monkey ganglion cells. J. Physiol., 199, 533-547.

Hawken, M.J. and Parker, A.J. (1984). Contrast sensitivity and orientation selectivity in lamina IV of the striate cortex of Old World monkeys. Exptl. Br. Res., 54, 367-372.

Hubel, D.H. and Wiesel, T.N. (1972). Laminar and columnar distribution of geniculo-cortical fibers in the macaque monkey. J. Comp. Neurol., 146, 421-450.

Livingstone, M.S. and Hubel, D.H. (1984). Anatomy and physiology of a color system in the primate visual cortex. J. Neurosci., 4, 309-356.

Lund, J.S. and Boothe, R.G. (1975). Interlaminar connections and pyramidal neuron organization in the visual cortex, area 17, of the macaque monkey. J. Comp. Neurol., 159, 305-334.

Lund, J.S., Fitzpatrick, D. and Blasdel, G.G. (1983). Intrinsic connections of macaque monkey striate cortex. Soc. Neurosci. Abstr., 9, 476.

Michael, C.R. (1978). Color vision mechanisms in monkey striate cortex: dual-opponent cells with concentric receptive fields. J. Neurophysiol., 41, 572-588.

Wiesel, T.N. and Hubel, D.H. (1966). Spatial and chromatic interactions in the lateral geniculate body of the rhesus monkey. J. Neurophysiol., 29, 1115-1156.

Colour Pathways and Colour Vision in the Honeybee

R. Menzel

Institut für Tierphysiologie und Angewandte Zoologie – Neurobiologie – Freie Universität Berlin, Königin-Luisestrasse 28–30, 1000 Berlin 33, Federal Republic of Germany

A. Introduction

Colour vision is a common receptor and neural mechanism in highly evolved invertebrates (for rev. see Menzel, 1979). As in many vertebrates the chromatic contrast of natural objects is evaluated besides the brightness contrast and used for visual orientation. However, the sensory and neural strategy in colour coding and the actual significance of colour dimensions might well be very different in animals which differ enormously in the organization of their eyes and nervous systems and whose colour vision has been shaped during evolution in very different ecological frameworks. Wavelengths specific behaviours for example, play a greater role in invertebrates than in vertebrates. These are visually guided behaviours, which are selectively under the control of a certain wavelength region (see Menzel, 1979, for references). Furthermore, wavelength selective eyes or eye regions are common features of invertebrate visual systems, e.g. pure green sensitive secondary eyes and trichromatic or tetrachromatic principle eyes in arachnid spiders (DeVoe, 1972, 1975, Yamashita and Tateda, 1976, Land, 1969), pure UV sensitive dorsal eyes and green sensitive ventral eyes in certain insect species (e.g. Ascalaphus, Bibio, Anax, drone bee, more examples in Menzel, 1979 pp 533ff). Chromatic separation at the level of the retina and at the level of the neural circuitry for specific behavioral patterns is an important strategy in invertebrates besides or in addition to colour vision.

In the honeybee colour vision and wavelength specific behaviors exist side by side. Fig. 1 gives a few examples of wavelength specific behaviors in bees. Polarized light orientation is monochromatic and restricted to the ultraviolet (300-400 nm) (v. Helversen and Edrich, 1974, Brines and Gould, 1979), whereas sun compas orientation is controlled by light 400 nm (Edrich, 1977, Brines and Gould, 1979). Large field motion detection is achromatic and dominated by the green receptors (Kaiser and Liske, 1974, Menzel, 1973). Phototaxis is either highly dominated by UV light in the case of experimentally induced escape phototaxis to strong lights or it is controlled equally well, by all wavelengths between 340 and 590 nm. Learning of spectral colours as signals for a food source is best for violet (400 nm) light (Menzel, 1967).

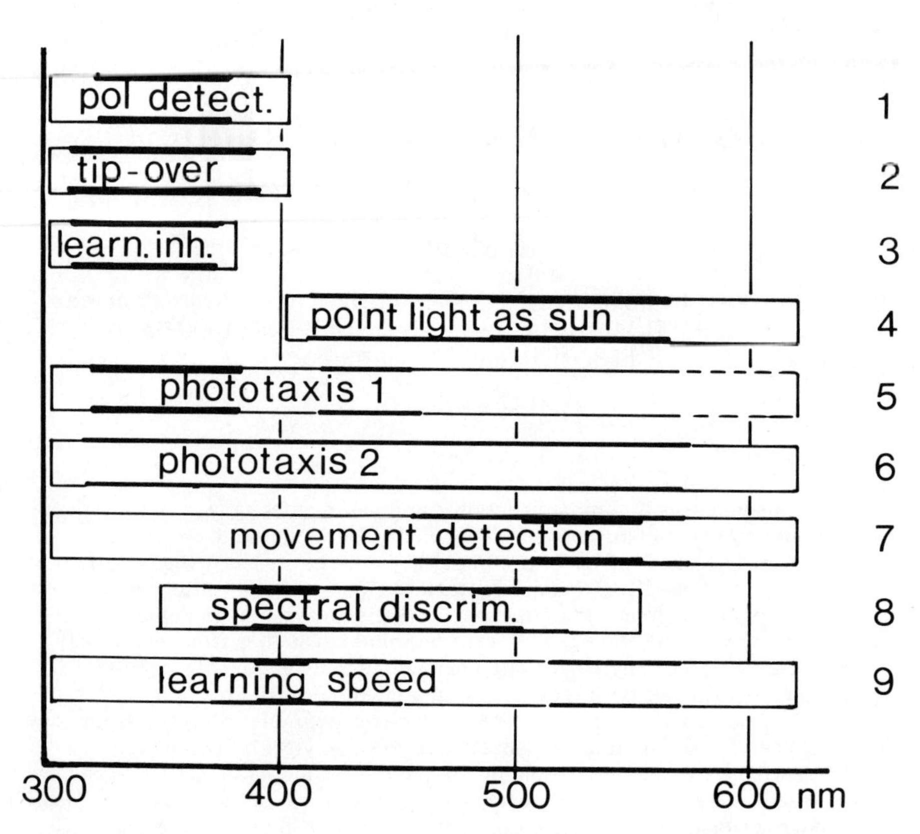

Fig. 1: Wavelength selective behaviour (1-7) and spectral dependencies of two behaviours including colour vision (8, 9). Three behaviours are restricted tipping over to monochromatic light illuminating a horizontal ground glass (2), reduction of the learning behavior of monochromatic lights at high light intensities (3). A point light source is interpreted as the sun if it excludes UV . Green light is more effective than other wavelength (4). Phototaxis is either highly dominated by UV-light in the case of an phototactic escape behaviour to strong light (5), or equally controlled by a large range of wavelengths between 340 and 580 nm in the case of natural, low light phototaxis (6). Movement detection to large field moving objects is achromatic and most sensitive to green light(7). The two behaviours including colour vision (8,9) are discussed in chapt. B and F.

Colour vision is used by bees in goal directed behaviour, e.g. in food selection and at the entrance to the hive (v. Frisch, 1914, 1967). Bees collect nectar and pollen as food from flowers. The colours of flowers are made for flower visiting insects like the honeybee, and are the result of a 300 million year long co-evolution with the visual system of these flower visiting insects. Many angiosperm plants need the insects as shuttle cars for the transportation of their genetic material. Obviously the transportation system works best for the plants, if the transporting vehicles only stop at flowers of the same species, loading and unloading only one kind of pollen. Many insects, particularly those which pollinate flowers, are fast flyers. Form vision is not an outstanding capability of the insects' compound eyes because of the physical constraints put on each of the little eyes (Kirschfeld, 1976). Colours are then the ultimate way of increasing contrast against the predominantly green background of the leaves. The green colour of chlorophyll is not only the background colour but also the colour of token marking the entrance to the hive. It would be, therefore, of great advantage if bees would be particularly sensitive to shades of green besides their sensitivity to those colours which contrast particularly well to green. This is indeed the case (see below).

B) Early behavioural studies on the honeybee's colour vision

Intuitively many biologists of the last century believed strongly that flower visiting insects like the bee see colours, since for what other reason should flowers be so brilliantly coloured (Sprengel, 1793, Darwin, 1877)? But there was no evidence, and later it was even denied on the basis of experiments with bees. v. Hess (1913) had bees flying towards self-luminant coloured lights and found that any preference for a particular wavelength could be changed by simply changing the intensity of the light. Bees perform phototactic flights under such conditions, and indeed they are colour blind in phototaxis (Menzel and Greggers, in press). However, v. Hess drew the wrong conclusion from his experiments that bees are generally colour blind. v. Frisch (1914) in his famous experiments demonstrated clearly that bees learn the colour blue as a marker for sucrose solution and discriminate the blue from any shade of grey (what was actually not really grey to bees but somewhat bluishgreen, see below). In these early experiments he also showed that bees are blind to red. Later he tested various other colours both at the feeding place and at the hive entrance (Ref. v. Frisch, 1967). Nowadays we know that bees discriminate many colours (see below) and that discrimination is similar at the feeding place and at the hive entrance (Menzel, in press).

The honeybees' sensitivity to ultraviolet light (300-400 nm) was proven by Kühn and Pohl (1921) and Kühn (1927) when they trained bees to parts of a spectrum produced by a quartz prism. Later Daumer (1956) established that bees are particularly sensitive to UV light around 350 nm. I shall come back to Daumer's most important experiments because he demonstrated the applicability of Graßmann's colour mixing rules to the bee's colour vision. Short and to the point Daumer found that UV (360 nm) and bluishgreen (490 nm) are complementary colours, as are blue (440 nm) and a mixture of UV and yellow (590 nm). He concluded, correctly as we know meanwhile, that bees have a trichromatic colour vision system with UV (360 nm), blue (440 nm) green (588 nm) being the primary colours.

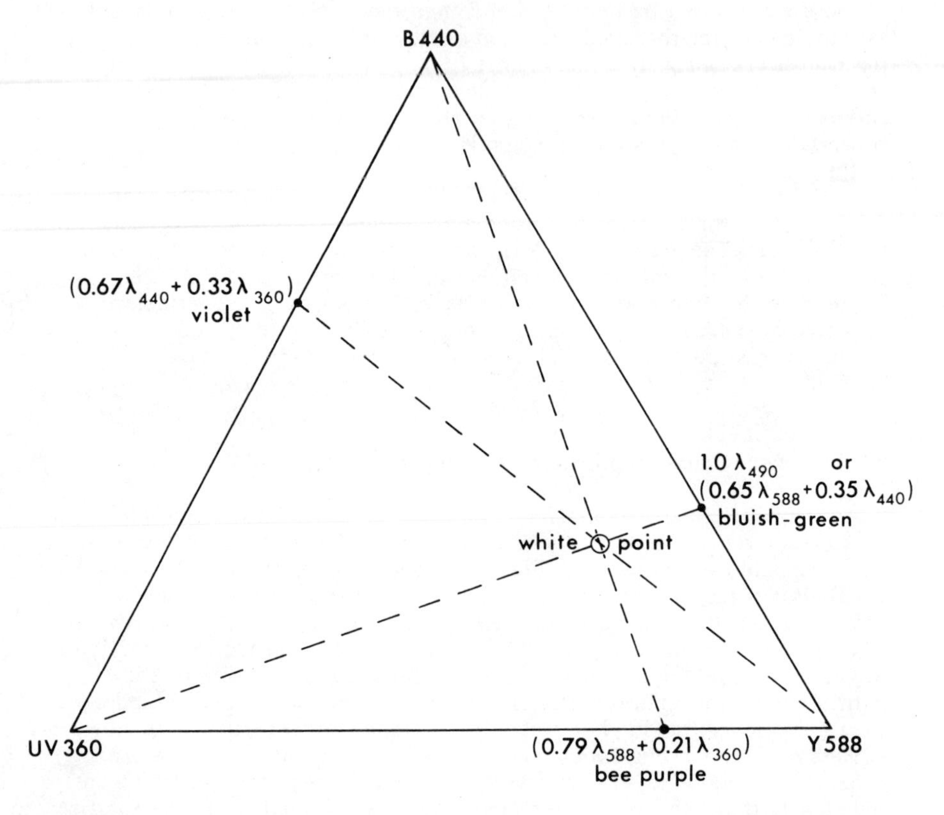

Fig. 2: The results of colour mixing experiments by Daumer (1956) (see text). The numbers are relative contributions of the wavelengths on energy basis.

Colour discrimination should be best, therefore, in the violet (around 400 nm) and bluishgreen (around 490 nm) region, and indeed this was found by v. Helversen (1972). Maximal discrimination values are 6 nm around 490 nm and 8 nm around 400 nm. In v. Helversen's experiments bees had to select a colour out of two alternative monochromatic colours by comparing each colour separately with the correct one stored in memory. One wonders whether colour discrimination would be even better if the test would allow a simultaneously comparison of two colours.

C) Colour receptors in the compound eye of the honeybee

Early in the history of intracellular recordings, the spectral sensitivities of individual photoreceptors in the bee eye were measured with intracellular electrodes (Autrum and v. Zwehl, 1964), and peak sensitivities in the UV (340-355 nm), in the blue (430, 460 nm) and in the green (540 nm) were found. With these results the honeybee was the first species

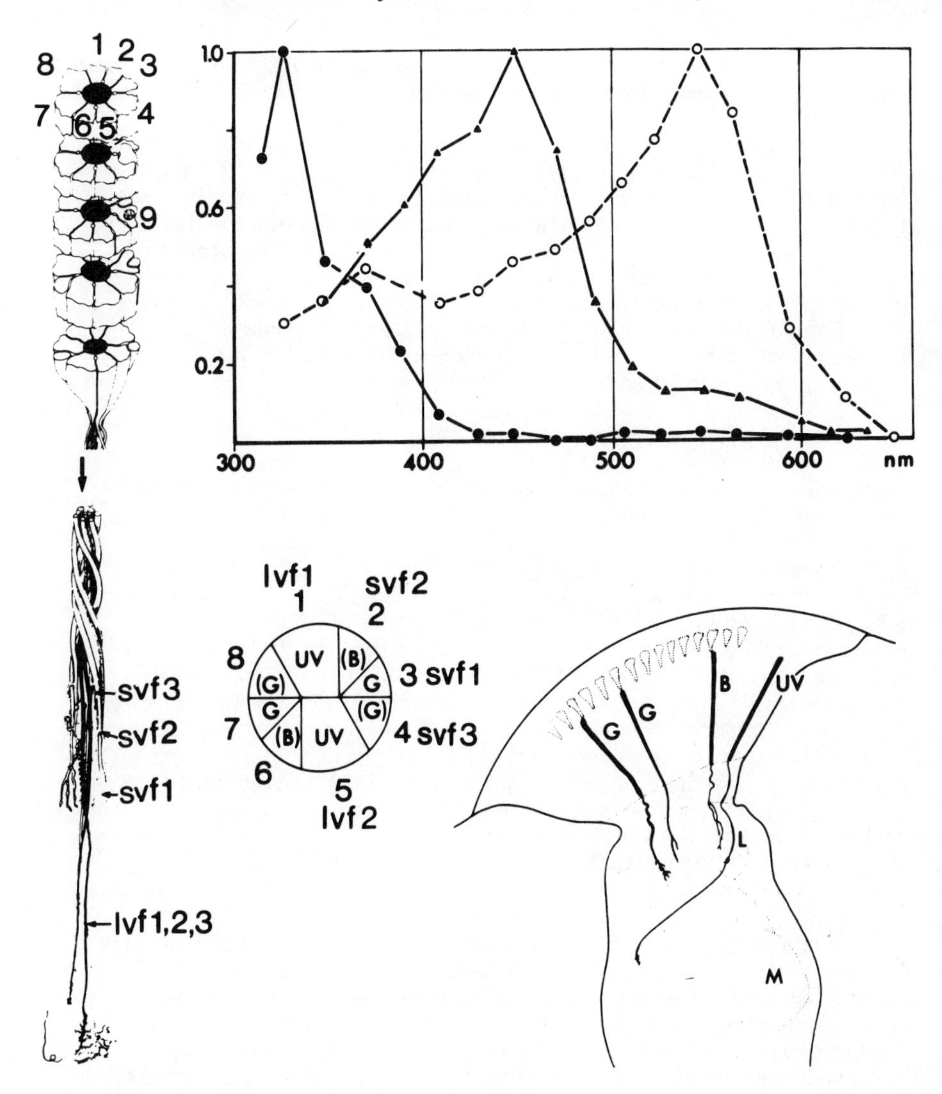

Fig. 3: Spectral sensitivity functions and localization of the 3 colour receptor types in median frontal ommatidia of the bee eye. The figure on the left side shows the proximal part of an ommatidium consisting of 9 receptor cells (No. 1-9). The axon of each receptor penetrates the basementmembran (at the level of the arrow) and projects to the lamina, the first optic neuropil (L in the right figure), or the medulla, the second optic neuropil (M in the right figure). Axons terminating in the lamina are called short visual fibres (svf 1-3), those extending to the medulla are called long visual fibres (lvf 1-3). The circular figure gives the arrangement of the 3 colour receptors UV, B and G in an ommatidium and relates their positions with the numbers of the retinula cells and their axons.

in which an inference on the chromatic properties of the input system from behavioural experiments were proven correct by intracellular recordings. These early recordings had a number of flaws, most likely due to recording artefacts. For example, pure UV receptors were found only in the drone bee; many spectral sensitivity functions had two or three peaks; two kinds of blue receptors, one with λ_{max} at 430 nm, another with λ_{max} at 460 nm, were found; stable recordings lasted not long enough to perform chromatic adaptation experiments. Further experiments were needed, therefore, to describe more accurately the spectral sensitivity functions and to work out the axonal projection pattern of the three receptor types.

Each of the nearly 5500 single eyes (called ommatidia or retinulae) in the compound eye of the bee contains 9 receptor cells (Phillips, 1905, Gribakin, 1970, Menzel and Snyder, 1974) (Fig. 3).

Eight photoreceptors are long, reaching down to the full depth of the retina, and one is short with its light absorbing structure, the rhabdomer, limited to the proximal quarter. This 9th cell is also a long cell in the most dorsal part of the eye (dorsal rim region, Schinz, 1975). Intracellular recordings and dye markings revealed that each of the median frontal ommatidia is composed of 4 green receptors (S ( ) with λ_{max} = 540 nm), 2 UV (λ_{max} = 350 nm) and 2 blue (λ_{max} = 440 nm) receptors (Menzel and Blakers, 1976 (Fig. 3). The short ninth cell has not yet been marked, but there is strong evidence that the 9th cell is a UV receptor (Menzel and Snyder, 1974). In the dorsal rim region this 9th cell is also most likely an UV receptor (Labhart, 1980). The light absorbing structures of all the retinula cells are fused into one homogeneous light guiding structure, the rhabdom. Since the 3 different photopigments are merged in one light guide they act on each other as spectral filters. This results in the narrow spectral sensitivity functions of single receptors resembling closely the spectral absorption of the respective photopigment in a thin layer, although total absorption is very high (90%; Snyder et al., 1973) (Fig. 4).

In summary, what we have learned from the intracellular markings of the spectral receptor types is that each median frontal ommatidium consists out of 4 green receptors, 2 blue receptors and 2 + 1 UV receptors. As each ommatidium looks to one point in space, the visual mosaic of the bee eye is simultaneously trichromatic in each point of view.

Measurements of the spectral sensitivities in the bee's photoreceptors by intracellularly placed electrodes revealed a large variability of the individual functions. This is not a peculiarity of the bee retina, but has been found in many arthropod eyes. There are sometimes much broader and sometimes narrower functions besides those which nicely resemble rhodopsin based functions. Many parameters have been claimed to influence the spectral sensitivity of a photoreceptor within the compound eye (see Menzel, 1974, 1979), for example the spectral transmission of the screening pigment; the wavelength dependent wave guide properties; the self-absorption of photopigments; the colour filters produced by stable reaction products of rhodopsin; the electrical coupling between retinula cells directly or through a return-current via the synaptic region; etc. Recently we have studied this question again and determined the

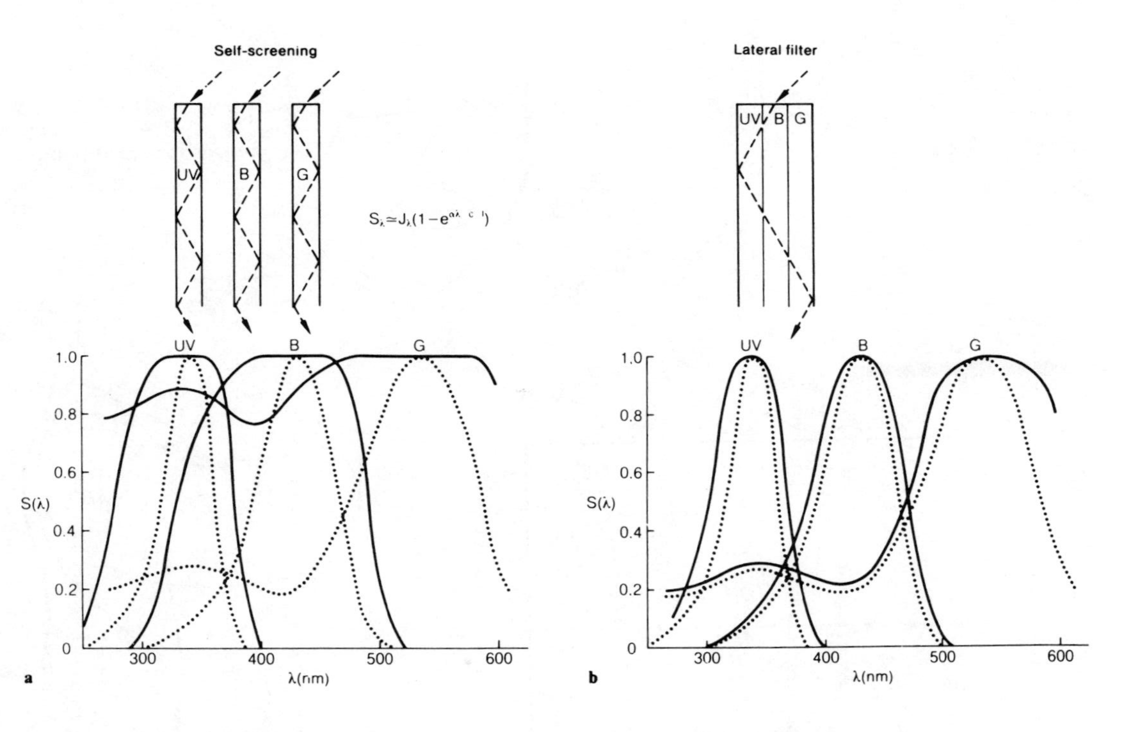

Fig. 4: The effect of lateral filtering in a fused rhabdom, which consists of sectors of photopigments absorbing preferentially in the UV, blue and green. a) The effect of high total absorption (c•l high as is the case in insect photoreceptors) in long isolated rhabdomeres results in a self-screening of the photopigment and thus in a flattening of the S (λ) function. b) The lateral filtering shapes the spectral sensitivity to nearly the functions of short rhabdomeres (c•l small) without self-screening.

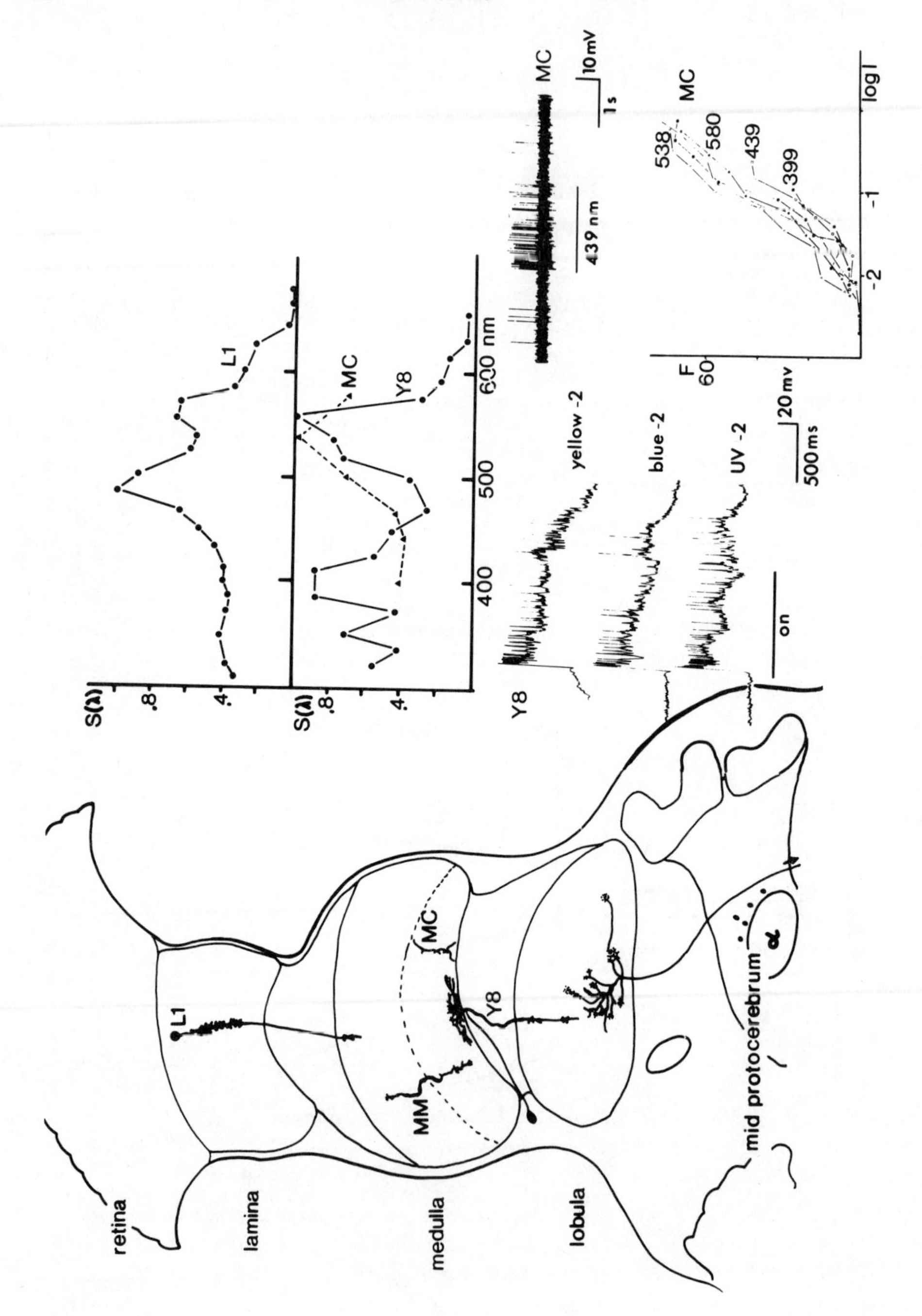
retina
lamina
medulla
lobula
mid protocerebrum
L1
MC
MM
Y8
S(λ)
.8
.4
400
500
600 nm
yellow -2
blue -2
UV -2
on
500 ms
20 mV
439 nm
1 s
10 mV
538
580
439
399
F
60
-2
-1
log I

lack of long time stability of the photoreceptor's sensitivity during intracellular recordings as the main source of deviation from a single peaked normal photopigment function (Menzel et al., in press). In contrast to flies and locusts intracellular recordings from the bee eye are not easy, and even if an electrode is placed successfully in a cell and large potentials ($>$20 mV peak potentials) are recorded on a stable baseline, the sensitivity of the cells changes considerably throughout a period of several minutes. These sensitivity changes are strong enough to alter substantially its spectral sensitivity if the traditional flash-method is used, which needs 3 mins or more to establish a spectral function. Much less variability is found with the voltage-clamp technique (Franceschini, see this vol.), which needs only 10-15 secs for a spectral run (see Fig. 9). We conclude from these new findings that relatively fast (10-15 sec) spectral measurements are needed to collect reliable data with the necessary spectral resolution. Furthermore, these new data reconfirm the 3 spectral classes of photoreceptors in the bee eye described earlier, and add the important result that antennal pigments like those in the fly (see Franceschini, this vol.) do not exist in the bee eye.

D) Luminosity coding

Neural colour coding is a mechanism to extract chromaticity differences independently of brightness differences. One of several theoretical possibilities is to encode the brightness of an object independently and separately from its chromaticity. In this case brightness should be measured over the whole wavelength range of vision, and result in a luminosity signal within the nervous system. The luminosity signal can then be used both as the input to a highly sensitive achromatic contrast detection network and as a reference for the colour coding system.

At the receptor level, particularly the broad green receptors would be candidates for a separate luminosity input. The two green receptors in each ommatidium with the thick, long axons (Fig. 5, No. 3, 7) may

Fig. 5: Luminosity coding neurons at various levels of visual integration (see text). L 1 is a first order interneuron in the lamina, MC a neuron intrinsic to the medulla and limited to a medullary column, MM another intrinsic medulla neuron, which is motion sensitive (no preferred direction, small receptive field). Neuron Y 8 connects the proximal medulla and the distal lobula. A lobula output neuron with a large receptive field is crossing from one side of the brain to the other (see arrow). The dots around the α-lobe of the mushroom body in the mid protocerebrum (•) mark recording sites of broadband neurons which respond to a certain range of intensity. Spectral sensitivity functions and responses are shown for a few selected examples on the right side. The response/intensity functions to various wavelengths is given for the medullary intrinsic neuron MC. Such R/logI functions are typical for broadband neurons in the visual ganglia.

be candidates, since their thicker axons and larger number of synaptic sites (Ribi, 1981) should improve the signal-to-noise ratio at the synaptic level. However, the S (λ) functions of these receptors are not broader than that of other green receptors. Further more, the pattern of synaptic contacts of these receptors is contraintuitive to the notion of broadband receptors (see Ribi, 1981 and below). Since we have not found any closer correlation of the sideband sensitivity or width of S (λ) in green receptors with the morphological type of receptors we believe that in contrast to flies and other insects, honeybees have no specialised luminosity receptors. But the question still remains why each ommatidium has two pairs of anatomically very different green receptors.

The first clear evidence for an achromatic broadband signal appears at the level of the lamina. Ribi (1981) analysed the synaptology in the lamina and found that one of the 4 first order interneurons in each cartridge, the lamina monopolar cell L 1, receives input from all receptors of the corresponding ommatidium (perhaps excluding the 9th receptor). (The other lamina monopolar cells are wired differently: L 2 receives input from all short visual fibres and L 3 only from svf 1, the fat green axons). Recordings which come most likely from neuron L 1 demonstrate a high sensitivity throughout the visual spectrum with a peak at the overlap of green and blue receptors (Fig. 5). Recordings from another lamina monopolar cell, perhaps L 2, show a spectral sensitivity restricted to the long wave part above 400 nm with a maximum in the bluishgreen (Menzel, 1974).

Nothing is known so far about the connectivity of the L 1 neuron within the next optic ganglion, the medulla, and recordings within this densely packed neuropile with its thousands of very thin fibres are extremely difficult to obtain. Recordings in the more proximal layers of the medulla and further into the brain show that there are broadband neurons at all levels of visual integration (Fig. 5).

In the medulla, these neurons have small receptive fields (e.g. 10°, see Hertel, 1980, his Fig. 1) when restricted to the medullary columns (MC in Fig. 5), or medium large receptive fields (60°, neuron Y 8, Fig. 5) when input is collected over a number of columns. Neurons with very large receptive fields ($> 60^{o}$) are found in the lobula and mid-protocerebrum (Kien and Menzel, 1977a, Hertel, 1980, Erber and Menzel, 1977). In particular, lobula output neurons connecting the contralateral optic ganglia, have very large receptive fields (Fig. 5). The receptive fields have not yet been analysed carefully enough, particularly not at different levels of light adaptation. So far medulla neurons with nearly circular, medium large receptive fields have sharp borders and are not subdivided in excitatory or inhibitory areas. Other neurons in the proximal medulla showed spatial antagonism but not in a centre surround fashion. Receptive fields may be subdivided into dorso-ventral stripes, one giving an excitatory the other an inhibitory input (Fig. 10 in Hertel, 1980). More central neurons display even more complicated receptive fields with different field sizes for different colours (see Fig. 10 in Kien and Menzel, 1977a), but no evidence is found for a centre-surround spatial antagonism.

The responses of broadband neurons were either sustained excitatory or inhibitory with or without additional ON and OFF excitations, or - more rarely - only phasic ON and OFF responses. Many of these neurons have very steep response/intensity functions (e.g. 0,5 log unit from threshhold to maximal response), others have more shallow R/logI functions (2-3 log I range), and some neurons, particularly in the lobula and mid-protocerebrum have rising and falling parts of their R/log I functions (see Erber and Menzel, 1977), so responding most effectively only to a certain intensity range. Neurons with ON and OFF transients are frequently sensitive to movement with or without prefered directions. Motion sensitive neurons in the medulla were direction insensitive with small or large receptive fields (e.g. neuron MM in Fig. 5), those in the lobula were always direction sensitive with large receptive fields (Hertel, 1980). All neurons sensitive to movement turned out to be spectrally broadband neurons without any indication of colour contrast effects. One of the large field motion sensitive neurons has been recorded and marked many times (Schäfer, 1984). This neuron receives input from only one eye at the level of the medulla. It responds best to a stripe (optimal size: 45° dorso-ventral, 11° horizontal) which is moved horizontally. The response increases steeply with the intensity of monochromatic light, reaching the highest response within 1-2 log I units from threshold. It has broadband sensitivity with a maximum in the green.

We conclude from our recordings that (1) luminosity signals are available at all levels of visual integration in the bee brain, (2) that all 3 receptor types contribute to the luminosity signal through a summation stage at the level of the first order interneuron (L 1), (3) that this luminosity signal serves an achromatic movement coding system with high spatial-temporal resolution and a stationary luminosity coding system with a very coarse spatial resolution but a fine intensity contrast resolution.

E) Colour antagonistic and narrowband neurons

Colour antagonism is the neural strategy of colour coding in the bee brain (Fig. 6). We find two major groups of colour antagonisms, one in which the sustained responses to flashes of monochromatic light is excitatory or inhibitory depending on the wavelength, and the ON-response is always excitatory for all wavelengths (tonic colour antagonistic neurons). The other group of neurons respond only with a short burst at ON or/and OFF to different monochromatic light stimuli (phasic colour antagonistic neurons). The tonic colour antagonism appeared in nearly all possible kinds of combinations, e.g. namely $UV^{+}B^{-}G^{-}$, $UV^{-}B^{+}G^{+}$, $UV^{-}B^{+}G^{-}$, UV^{+}, B^{-}, G^{+}. Combinations not found are: $UV^{+}B^{+}G^{-}$, $UV^{-}B^{-}G^{+}$. Phasic colour antagonism is found for OFF-response to UV, ON-response to blue, OFF-response to green light (UV-OFF, B-ON, G-OFF, see Fig. 6 lowest trace), and no other responses (Hertel, 1980). In both types of colour antagonism the excitatory or inhibitory response increases with the intensity of the monochromatic light stimulus. This allows us to calculate sensitivity functions for both the excitatory and inhibitory spectral regions (e.g. Fig. 6 for $UV^{+}B^{-}G^{-}$ tonic colour antagonistic neurons). One can easily read from such spectral sensitivity

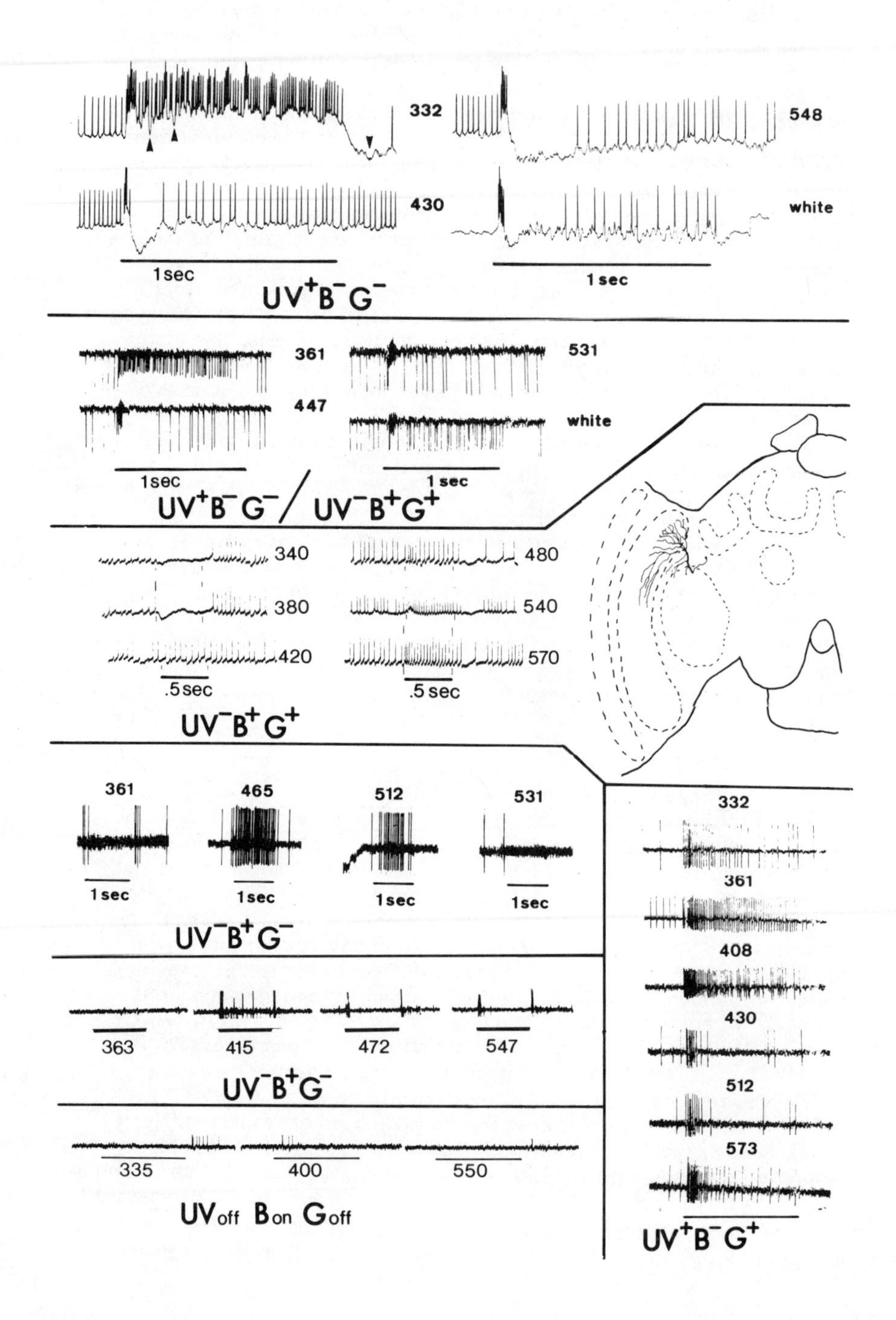
332
548
430
white
1sec
1 sec
UV+B−G−
361
531
447
white
1sec
1 sec
UV+B−G−/ UV−B+G+
340
480
380
540
420
570
.5sec
.5 sec
UV−B+G+
361
465
512
531
1 sec
UV−B+G−
363
415
472
547
UV−B+G−
335
400
550
UVoff Bon Goff
332
361
408
430
512
573
UV+B−G+

functions how the 3 receptor types contribute to the chromatic sensitivities of these neurons (see Kien and Menzel, 1977b, Hertel, 1980).

In other neurons the inhibitory components are not so obvious because of the low rate of spontaneous discharge. These neurons have frequently very narrow spectral sensitivity functions (Fig. 7). We found spectral peaks not only at the λ_{max} of the 3 colour receptor types but most interestingly also at the spectral regions where the 3 receptors overlap (Fig. 7).

The intensity dependence may sometimes be very complicated. For example a neuron in the proximal layer of the medulla responded excitatoryly to UV and inhibitoryly to green at a medium intensity and in a reversed fashion at higher intensities Fig. 6 in Hertel, 1980). Furthermore, neurons may respond differently at different regions of their receptive field, e.g. with an OFF burst to green light in the median-ventral eye region and with ON/OFF bursts to UV light in the dorsal eye region (Fig. 8 in Hertel, 1980).

We have learned from results of this kind, that colour antagonism is an important feature of the neural code of colour in the bee brain, but nothing is known so far about the sequence of colour antagonistic processing in the visual ganglia.

Furthermore, too little information has been gathered so far on the combination of colour and spatial sensitivity. The only rule so far established is that there are no concentric double antagonistic (colour, spatial) receptive fields. Many neurons have complicated receptive fields with bar and dot like substructures, which differ with respect to their optimal colour. The impression one gets from these neurons is that there may be a set of individual neurons with very specific combinations of space and colour properties. It would be fascinating to know the level of visual integration to which such neurons belong, how these complicated features are established and whether these neurons change their properties with experience. Such studies seem to be feasible with the honeybee.

Fig. 6: A summary of all kinds of colour antagonistic neurons recorded in the visual ganglia of the bee (see text). Except of the very lowest trace all neurons display tonic colour antagonism. The lowest trace gives an example for a phasic colour antagonistic neuron (UV-OFF, B-ON, G-OFF). The neuron UV^{-} B^{+} G^{+} in the middle of the figure has been intracellularly marked. It receives input from the proximal medulla and projects to the distal dorsal lobula (Schäfer, 1984).

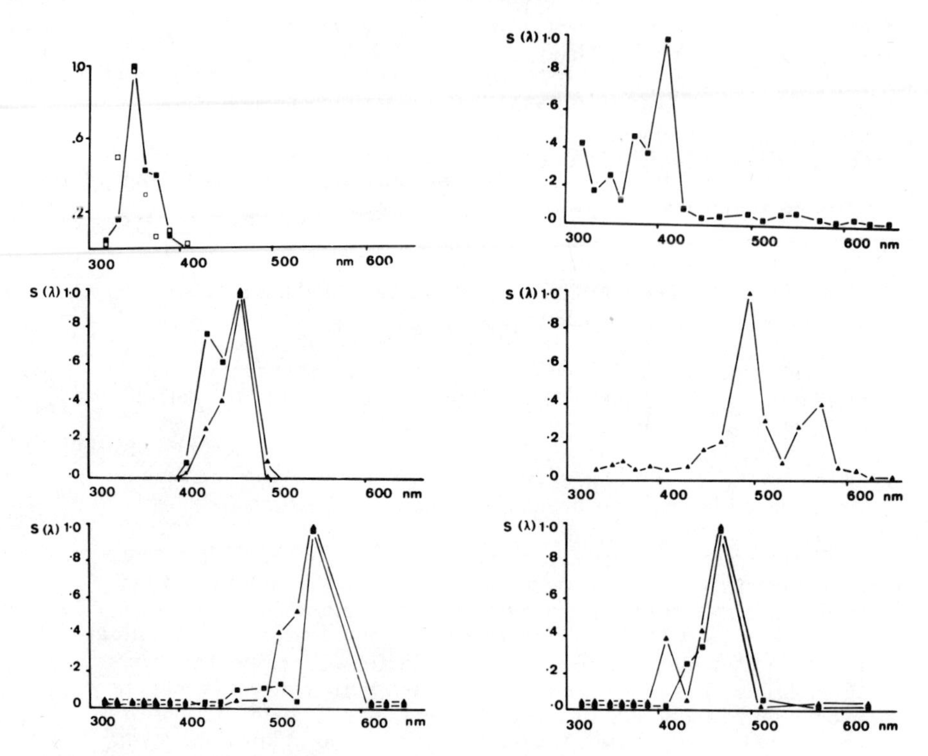

Fig. 7: Narrowband colour coded neurons in the lobula region of the bee brain. The 3 examples on the left side resemble the chromatic properties of the 3 colour receptor inputs but with narrower spectral sensitivities. The 2 examples on the right side have their maximal spectral sensitivities at wavelength regions (violet, bluishgreen) were the colour receptors overlap.

F) Colour vision: Recent behavioural studies

Daumer (1956) concluded from his colour mixing experiments that colour vision in man and bee is in principle similar with the only difference that the visual range is shifted by 100 nm to shorter waves in the bee. He applied Schrödinger's (1924) formalism to the bee and came up with the following scheme:

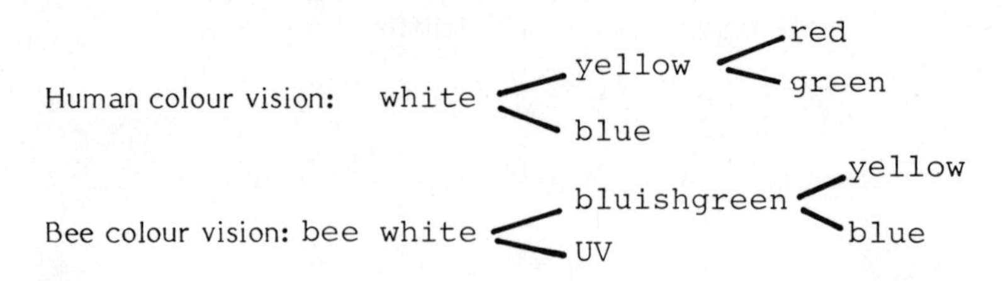

Such a comparison formulates predictions for the psychophysical dimensions of colours in bees, which have to be tested with the relevant psychophysical methods. Although trichromacy is realized at the input in bees, too, and colour antagonism is also a basic feature in colour coding, there is no a priori constraint to assume that bees have the same 3 perceptual dimensions (hue, saturation and brightness) as humans, that bees experience unique hues, and that bees enjoy the phenomena related to colour constancy, etc. There are no affirmative answers to these questions yet, but research in this area is active and has come up with relevant information. This shall be pointed out in the following paragraphs.

- Achromatic interval

Bees can be trained to make a right turn in a T-maze when the intersection is illuminated with white light and a left turn when the intersection is illuminated with a monochromatic light (Menzel, 1981). If the intersection is illuminated with the monochromatic light at various intensities, the bee behaves in a certain range of intensity as if the intersection is illuminated with achromatic white light (Fig. 8). Above that intensity range the behaviour is correct, that means bees choose that arm to which they were trained in response to the coloured light, below that intensity range, the bees turn into the arm to which they were trained in response to the white light, and at even lower intensities they choose the two arms randomly.

I interpreted these results as evidence for two thresholds, a higher one for colour vision, a lower one for light detection. The intensity range between the two thresholds may be called achromatic interval (see Bouman and Walraven, 1957, Massof, 1977). The wavelength dependence of the achromatic interval has not yet been worked out carefully enough. The few colours tested gave about the same result (1,5 log I). A tentative conclusion is that the achromatic signal results from a pooling of all 3 colour receptor inputs. This is in agreement with our conclusion that all 3 colour receptors contribute to the luminosity signal.

Under natural light conditions bees discriminate colours as long as they fly. If bees are manipulated to fly at lower light intensities they behave colour blind and approach any contrasting object irrespective of its colour (Rose and Menzel, 1981). The light fluxes at threshold of colour vision correspond nicely in the two experiments. We calculated an average of about 5 - 50 quanta per facet lens and integration time (30 msec) at absolut threshold and a value of about 100 - 1000 quanta per facet lens and integration time at colour vision threshold.

- Spectral threshold function and $\Delta\lambda$-function

Spectral threshold has been determined in colour training experiments several times (Daumer, 1956, v. Helversen, 1972, Menzel, 1967, 1981, Lieke, pers.com.). If the background light excludes UV, the sensitivity to UV can be as much as 10 times higher than to blue and green light, but if there is some UV in the background light, the UV sensitivity is only slightly enhanced. Dark adapted bees tested in the dark, are about equally sensitive to UV and other wavelengths.

v. Helversen (1972) determined the $\Delta\lambda$ - function for spectral lights with freely flying bees at low background light with little UV. He found optimal colour discrimination in the violet (400 nm : $\Delta\lambda$ = 8 nm) and in the bluishgreen (490 nm : $\Delta\lambda$ = 6 nm). Discrimination is less than 30 nm for wavelengths $\leq$ 360 nm and $>$ 530 nm; in blue (440 nm) $\Delta\lambda$ = 16 nm. It has been pointed out above (see Chapt. 1) that the bee had to compare a test colour with the memory of the correct colour in v. Helversen's experiments. We expect an improved colour discrimination in experiments which allow a simultaneous comparison with the test colours presented within the same field of view. Such experiments are in progress.

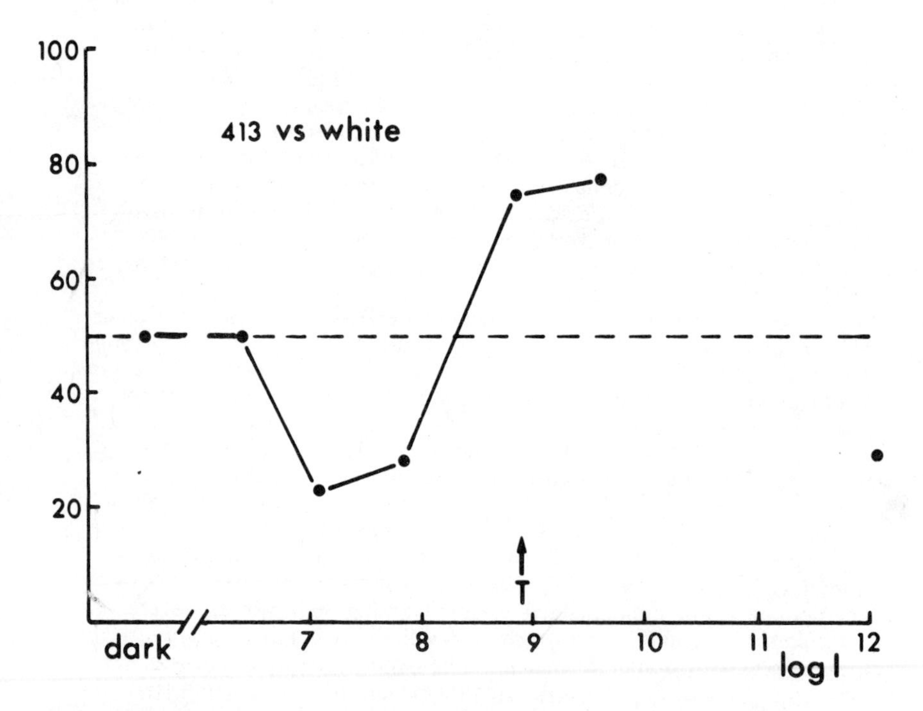

Fig. 8: Achromatic interval. Dark adapted bees were trained in a T-maze to turn to one side when the intersection was illuminated with monochromatic light of 413 nm and to the other side when it was illuminated with white light of the same subjective brightness. T gives the quantal flux (log quanta cm^{-2} sec^{-1}) of 413 nm during training. In the test situation the monochromatic light was presented at various intensities (abscissa) and the response (ordinate) observed. Responses 50% are correct with respect to the training to the monochromatic light, responses 50% are correct with respect to white light (see dot at the right corner: response level to white light). Responses at 50% reflect random choices of the two arms of the T-maze. All points above or below 50% are significantly different from the responses at 50%. Dark: intersection not illuminated.

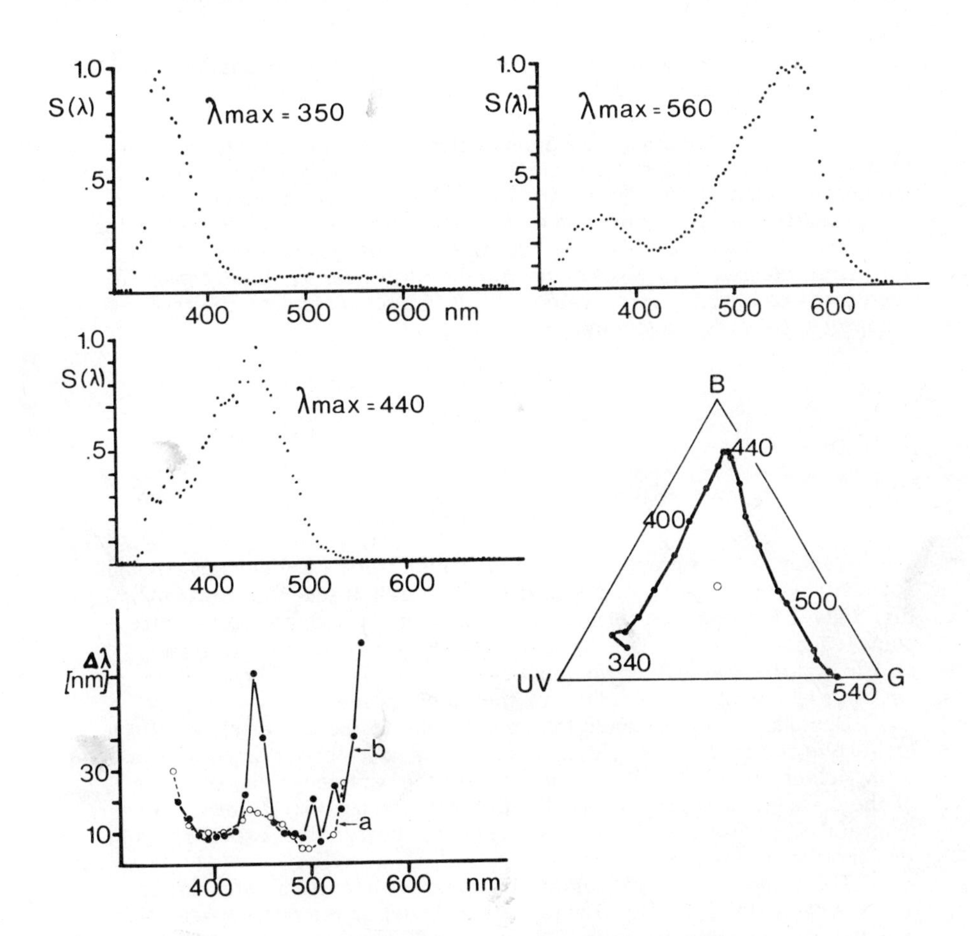

Fig. 9: Our most recent measurements of the S (λ) functions of the 3 colour receptor types are used to calculate a receptor-based chromaticity diagram. The S (λ) function of a receptor is measured by the voltage clamp technique (see text) within less than 20 sec (see text) and with a wavelength resolution of 4 nm. The functions shown here are averages of n = 39 cells with 128 runs (for the green receptors), n = 5 cells with 18 runs (for the blue receptors), and n =10 cells with 27 runs (for the UV receptors) (Menzel and de Souza, unpubl.). The chromaticity diagram is calculated by the formulas given in the text. The diagram at the lower left side gives the Δλ- function measured by v. Helversen (1972) (curve a) and a normalized function of the distance between neighbouring loci of spectral lights along the line for spectral colours (curve b). Colour discrimination is better in blue and in bluishgreen than expected from the distance between the colour loci, but the wavelengths of optimal colour discrimination (around 400 nm and 490 nm) coincide very well.

- A receptor based chromaticity diagram of the bee's colour space

Colour-mixture functions have not yet been established for the bee, although a few numbers are known from Daumer's (1956) experiments. Therefore, it is not possible to calculate a proper chromaticity diagram from the spectral tristimulus values as in humans (CIE chromaticity diagram). On the other side, the spectral properties of the receptors are known very well, and they can be used to calculate a receptor based chromaticity diagram by making a few reasonable assumptions. Following Rushton (1972) we normalize the average relative S (λ) functions of the 3 receptor classes (Fig. 9) such that the integral over wavelengths of each function become unity (values UV, B, G), and calculate the tri-receptor values uv, b and g by:

$$uv = \frac{UV}{UV+B+G}, \quad b = \frac{B}{UV+B+G}, \quad g = \frac{G}{UV+B+G} \quad \text{and}$$

$$uv+b+g = 1.$$

The meanings of these operations in functional terms are obvious, though not justified by physiological results (for discussion see Rushton, 1972). Fig. 9 gives the result in form of a colour triangle with the ideal white point (equal numbers of light quanta over the whole spectrum) in the middle and the loci of the spectral colours from 340 to 560 nm in 10 nm intervals along the periphery of the colour plane.

Obviously colours should look more alike to the bee as closer their loci are in the colour triangle, and indeed the $\Delta\lambda$- function established by v. Helversen (1972) follows quite nicely the wavelength dependence of the distance between the loci of spectral colours (Fig. 9, lower left diagram). Discrimination in the blue region is better than expected from this simple formula.

The value of such a chromaticity diagram lies in its potential to arrange the colour of natural objects in a model of the perceptual space, which allows to estimate at least roughly, how similar colours look like to the bee, and thus allow to formulate hypothesis about the subjective appearence of colours. Indeed, we have calculated the colour loci of a large number of natural and arteficial coloured objects and found very good correspondence between the distance of the colour loci and the discrimination of the colours (Backhaus and Menzel, 1984, Werner and Menzel, 1984).

- Colour saturation

Evidence for saturation as a dimension in the bee's colour space comes from two very recent studies. Lieke (1984) trained bees to white light and tested discrimination between a pure white light and a white light to which various amounts of different spectral lights were added. He carefully choose the intensities in such a way that both alternatives, the pure white light and the mixed light, appeared equally bright to the bee judged from separately determined response/intensity functions. He found that bees discriminate the alternatives if more than 8% of monochromatic light is added to the white light. Most interestingly, the wavelength dependence is very different from the spectral threshold function, e.g. UV (360 nm) light is somewhat less effective than bluishgreen (490 nm) light in these colour mixing experiments, but more effective in spectral threshold experiments.

Backhaus and Menzel (1984) used a multidimensional analysis to interprete dual choice and multiple choice colour discrimination experiments with colour targets varying in λ_{max} and purity. Following Thurstone's (1927) law of comparative judgement the relative choice values were used as a measure of similarities between the coloured targets. These similarity values were interpreted as distances in a n-dimensional euclidian space. On that basis the vector space was determined with the lowest number of dimensions. Two vectors were found which correlate very well with the hue and saturation axis in a chromaticity diagram.

We conclude from both experiments that bees discriminate colours according to differences in spectral purity, and thus may have a dimension of colour saturation. The multidimensional scaling shows that hue differences are primarily used for discrimination and saturation differences become important when hue differences are very small.

- Colour contrast effects

Simultaneous and/or successive colour contrast effects were described by Neumeyer (1980). She had freely flying bees choosing between several coloured targets which differed in hue along the bluegreen axis, and tested whether the choice preference was changed according to the colour of the large field background or of the immediate surround. A separation between simultaneous and successive contrast was tried but was seriously impaired by the possibility to control for displacement of the coloured image on the moving eye. The compound of simultaneous and successive contrast, at least, is very pronounced: blue targets look more blue to the bee on yellow background and less blue on blue background, yellow targets appear more yellow on blue background and less yellow on yellow background. The same is true for narrow coloured rings surrounding the coloured targets. It is unknown whether the colour contrast induction effects change the hue of the whole target or whether they are limited to the boarders.

Long lasting colour adaptation effects were found, when bees were exposed to blue or yellow adaptation colour for several minutes. Blue targets look more blue to the bee within a period of about 1,5 min following a 5 min adaptation time to a yellow background, and vice versa. The long duration of the selective adaptation effect suggests strongly neuronal mechanisms but receptor adaptation may be involved, too.

- Colour constancy

Colour appearence seems to change very little or not at all for various natural and arteficial illuminations (Mazokhin-Porshnyakov, 1966, Neumeyer, 1981, Menzel, unpubl. results), but neither the range nor the chromatic dependencies of the constancy effects were investigated so far. Particularly, changes in UV illumination were not studied, although such changes are most important under natural conditions. Furthermore, nothing is known about the spatial properties of the constancy effects. It is reasonable to suppose that lateral inhibitory interactions are the main reasons in colour constancy because lateral inhibition has been shown to control the level of adaptation in the first order interneurons of insects (Laughlin, 1981).

- Combined colour - spatial antagonistic effects

Although spatial resolution is low in compound eyes, bees recognize very well patterns with spatial wavelengths above 1,5° (Wehner, 1981). Patterns made of colours with optimal colour contrast and excluding brightness contrast are discriminated equally well or even better than black and white patterns (Menzel and Lieke, 1983). The position of colours in a vertically arranged colour pattern is of significance for pattern discrimination. For example, a circular pattern with an UV half-field and a bluishgreen half-field is chosen most frequently, if the UV part is positioned in the upper half and the bluishgreen part in the lower part. A similar orange-blue pattern receives most choices if orange covers the upper half and blue the lower half. The questions related to learned and innate search images for coloured patterns are discussed in detail in Menzel (1984). What shall be pointed out here is that the colours in the two pairs of colours, UV-bluishgreen and blue-orange, are antagonistically related to each other with respect to the position in space. An explanation could be that visual neurons in the bee brain code for a limited number of spatial arrangements of colours. Colour patterns which fit best these specialized detectors may then be choosen more frequently and more accurately. The lack of circular receptive fields with centre/surround double opponency support this notion, since such neurons are least specialized and multi-functional. Specialized neurons have been recorded many times (see Chapt. 4) including those which are excited by UV illumination in the upper eye region and green illumination in the lower eye region.

- Colour values

Colours have meanings for the bee both as a consequence of individual experience and as the result of the phylogenetic memory inherited in the brain. It is reasonable to expect, therefore, that different colours have different values as sign-posts for visual orientation. The visual learning system of the bee is so flexible and general that every perceived colour it sees can be learned as marker at the food source or at the hive entrance. A quantative analysis of the learning performance indicates, however, that the values of certain colours are higher and others are lower. Violet (400 nm), is learned fastest; bees need only one reward to choose it more accurately than 85% in a dual choice situation and reach more than 95% correctness after more than 3 rewards. Bluishgreen instead is learned slowest; they need more than 5 rewards to reach 85% correctness and never get better. The other colours lie in between these extremes (Menzel, 1967). Violet and bluishgreen differ in no obvious way at the sensory side. Both colour regions are discriminated very well, appear about equally bright and add about the same in colouring a white light. The difference refers, therefore, to an evaluation process in the brain, which must be separate from the sensory coding and adds meaning to the sensory code.

CONCLUSION

There are not many animal species, whose colour vision system have been carefully analysed. The honey bee is one of the few. The fascination of working on colour vision in honey bees derives from the combination of neurophysiological and behavioural approaches. Although recordings from the interneurons are not easy and behavioural work is time consuming, the combination of both shall ultimately allow us to describe psychophysical terms with neurophysiological terminology.

References

1. Autrum, H. and v. Zwehl, V. (1964). Spektrale Empfindlichkeit einzelner Sehzellen des Bienenauges. Z. vergl. Physiol. 48, 357-384.
2. Backhaus, W. and Menzel, R. (1984). Bestimmungen der Farbwahrnehmungskomponenten bei Bienen durch multidimensionale Skalierung. Verh. Dtsch. Zool. Ges. Gustav Fischer Verlag, Stuttgart.
3. Bouman, M.A. and Walraven, P.L. (1972). On the threshold mechanisms for achromatic and chromatic vision. Acta Psychol. 36, 178-247.
4. Brines, M.L. and Gould, J.L. (1979). Bees have rules. Science 206, 571-573.
5. Darwin, Ch. (1877). The different forms of flowers on plants of the same species. John Murray, London.
6. Daumer, K. (1956). Reizmetrische Untersuchungen des Farbensehens der Bienen. Z. vergl. Physiol. 38, 413-478.
7. DeVoe, R.D. (1972). Dual sensitivity of cells in the wolf spider eyes at ultraviolet and visible wavelength of light. J. Gen. Physiol. 59, 247-269.
8. DeVoe, R.D. (1975). Ultraviolet and green receptors in principal eyes of jumping spiders. J. Gen. Physiol. 66, 193 ff.
9. Edrich, W. (1977). Die Rolle einzelner Farbrezeptortypen bei den verschiedenen Lichtreaktionen der Biene. Verh. Dtsch. Zool. Ges. 70, 236.
10. Franceschini, N. (1983). In-Vivo microspectrofluorimetry of visual pigments. In The Biology of Photoreception. (eds. D.J. Cosens and D. Vince-Price). Society for Exp. Biology, Great Britain.
11. Gribakin, F.G. (1979). The distribution of the long wave photoreceptors in the compound eye of the honeybee as revealed by selective osmic staining. Vision Res. 12, 1225-1230.
12. v. Helversen, O. (1972). Zur spektralen Unterschiedsempfindlichkeit der Honigbiene. J. Comp. Physiol. 80, 439-472.
13. v. Helversen, O. and Edrich, W. (1974). Der Polarisationsempfänger im Bienenauge: Ein Ultraviolettrezeptor. J. Comp. Physiol. 94, 33-47.

14. v. Hess, C. (1913). Experimentelle Untersuchungen über den angeblichen Farbensinn der Bienen. Zool. Jb. (Physiol.) 34, 81-106.
15. Kirschfeld, K. (1976). The resolution of lens and compound eyes. In Neural Principles of Vision. (eds. F. Zettler and R. Weiler). Springer, Berlin.
16. Kühn, A. (1927). Über den Farbensinn der Bienen. Z. vergl. Physiol. 5, 762-800.
17. Kühn, A. and Pohl, R. (1921). Dressurfähigkeit der Bienen auf Spektrallinien. Naturwissensch. 9, 738-740.
18. Labhart, T. (1980). Specialized photoreceptors at the dorsal rim of the honeybee's compound eye: polarizational and angular sensitivity. J. Comp. Physiol. 141, 19-30.
19. Land, M.F. (1969a). Structure of the retina of the principle eyes of jumping spiders (Salticidae, Dendryphantinae) in relation to visual optics. J. exp. Biol. 51, 443-470.
20. Laughlin, S.B. (1981). Neural principles in the visual system. In Handbook of Sensory Physiology Vol. VII/6B. (ed. H. Autrum). Springer, Berlin-Heidelberg-New York.
21. Lieke, E. (1984). Farbensehen bei Bienen: Wahrnehmung der Farbsättigung. Dissertation, Freie Universität, Berlin.
22. Massof, R.W. (1977). A quantum fluctuation model for foveal color thresholds. Vision Res. 17, 565-570. 23.Mazokhin-Porshnyakov, G.A. (1966). Recognition of coloured objects by insects. In The Functional Organization of the Compound Eye. (ed. S.G. Bernhard). Pergamon Press, Oxford.
24. Menzel, R. (1967). Das Erlernen von Spektralfarben durch die Honigbiene. Z. vergl. Physiol. 56, 22-62.
25. Menzel, R. (1973). Spectral response of moving detecting and "sustained" fibres in the optic lobe of the bee. J. Comp. Physiol. 82, 135-150.
26. Menzel, R. (1974). Color receptor in insects. In The compound Eye and Vision in Insects. (ed. G.A. Horridge). Clarendon Press, Oxford.
27. Menzel, R. (1979). Spectral sensitivity and colour vision in invertebrates. In Handbook of Sensory Physiology Vol. VII/6A. (ed. H. Autrum). Springer Verlag, Berlin-Heidelberg-New York.
28. Menzel, R. (1981). Achromatic vision in the honeybee at low light intensities. J. Comp. Physiol. 141, 389-393.
29. Menzel, R. (1984). Learning in honey bees in an ecological and behavioral context. In Experimental Behavioral Ecology and Sociobiology. (eds. B. Hölldobler, M. Lindauer). Fortschr. Zool. 31. Gustav Fischer Verlag, Stuttgart.
30. Menzel, R. and Blakers, M. (1976). Colour receptors in the bee eye - morphology and spectral sensitivity. J. Comp. Physiol. 108, 11-33.
31. Menzel, R. and Greggers, U. (in press). Natural phototaxis and its relation to color vision in honey bees. J. Comp. Physiol.
32. Menzel, R. and Lieke, E. (1983). Antagonistic color effects in spatial vision of honey bees. J. Comp. Physiol. 151, 441-448.
33. Menzel, R. and Snyder, A.W. (1974). Polarized light detection in the bee, Apis mellifera. J. Comp. Physiol. 88, 247-270.

34. Menzel, R., Ventura, D., Hertel, H. and de Souza, J. (submitted). Spectral measurements of individual photoreceptors in bees: a critical evaluation of the methods and results.
35. Neumeyer, Ch. (1980). Simultaneous color contrast in the honeybee. J. Comp. Physiol. 139, 165-176.
36. Neumeyer, Ch. (1981). Chromatic adaptation in the honeybee: Successive color contrast and color constancy. J. Comp. Physiol. 144, 543-553.
37. Phillips, E.F. (1905). Structure and development of the compound eye of the honeybee. Proc. Acad. Nat. Sci. (Philad.) 57, 123-157.
38. Ribi, W.A. (1981). The first optic ganglion of the bee. IV. Synaptic fine structure and connectivity patterns of receptor cell axons and first order interneurons. Cell Tissue Res. 215, 443-463.
39. Rose, R. and Menzel, R. (1981). Luminance dependence of pigment color discrimination in bees. J. Comp. Physiol. 141, 379-388.
40. Rushton, W.A.H. (1972). Pigments and signals in colour vision. J. Physiol. (Lond.) 220, 1-31.
41. Snyder, A.W., Menzel, R. and Laughlin, S.B. (1973). Structure and function of the fused rhabdom. J. Comp. Physiol. 87, 99-135.
42. Schäfer, S. (1984). Charakterisierung extrinsischer Grossfeldneuronen aus der Medulla der Honigbiene (Apis mellifera). Diplomarbeit, Freie Universität, Berlin.
43. Schinz, R.H. (1975). Structural specialization in the dorsal retina of the bee, Apis mellifera. Cell Tissue Res. 162, 23-34.
44. Schrödinger, E. (1924). Über den Ursprung der Empfindlichkeitskurven des Auges. Naturwissensch. 12, 927-935.
45. Sprengel, Chr.K. (1793). Das entdeckte Geheimnis der Natur im Bau und in der Befruchtung der Blumen. Berlin.
46. Thurstone, L.L. (1927). A law of comparative judgment. Psychol. Res. 34, 273.
47. Wehner, R. (1981). Spatial vision in Arthropods. In Handbook of Sensory Physiology Vol. VII/6C. Springer Verlag, Berlin-Heidelberg-New York.
48. Werner, A. and Menzel, R. (1984). Farbunterscheidung und Rezeptorfarbraum der Honigbiene. Verh. Dtsch. Zool. Ges., Gustav Fischer Verlag, Stuttgart.
49. Yamashita, S. and Tateda, H. (1976). Spectral sensitivities of jumping spider eyes. J. Comp. Physiol. 105, 29-41.

Index